Springer Textbooks in Earth Sciences,
Geography and Environment

The Springer Textbooks series publishes a broad portfolio of textbooks on Earth Sciences, Geography and Environmental Science. Springer textbooks provide comprehensive introductions as well as in-depth knowledge for advanced studies. A clear, reader-friendly layout and features such as end-of-chapter summaries, work examples, exercises, and glossaries help the reader to access the subject. Springer textbooks are essential for students, researchers and applied scientists.

More information about this series at http://www.springer.com/series/15201

Maurizio Petrelli

Introduction to Python in Earth Science Data Analysis

From Descriptive Statistics to Machine Learning

 Springer

Maurizio Petrelli
Department of Physics and Geology
University of Perugia
Perugia, Italy

ISSN 2510-1307   ISSN 2510-1315 (electronic)
Springer Textbooks in Earth Sciences, Geography and Environment
ISBN 978-3-030-78054-8   ISBN 978-3-030-78055-5 (eBook)
https://doi.org/10.1007/978-3-030-78055-5

This Springer imprint is published by the registered company Springer Nature Switzerland AG
The registered company address is: Gewerbestrasse 11, 6330 Cham, Switzerland

To my daughters Agata, Anna, and Caterina, my wife Arianna, and Atomo who completes the family

Preface

The idea of writing this book came to me in 2015 when I started teaching a course entitled "Data Analysis and Interpretation in Earth Science" at the Department of Physics and Geology of Perugia University. From the beginning of the course, I realized that many of my students were strongly interested in data managing, visualizing, and modeling in Python. I also realized that no reference book was available for teaching Python to geologists. Although numerous books present Python to programmers at all levels, from beginners to experts, they mostly focus solely on programming techniques, without discussing real applications, especially in geology. In other words, a book devoted to Earth Scientists was missing. The project grew and became structured while teaching the basics of Python to Earth Scientists at the Eötvös University Budapest (Hungary) and at Leibniz Universität of Hannover (Germany) in December 2018 and February 2020, respectively. Sadly, by the beginning of March 2020, the COVID-19 pandemic had dramatically spread to all regions of Italy and, on March 4th, the Italian government shut down all schools and universities nationwide, forcing me to stay at home like most Italians. In one of the most confusing and insecure moments of my life, I decided to start writing this book. "Introduction to Python in Earth Science Data Analysis" is devoted to Earth Scientists, at any level, from students to academics and professionals, who would like to harness the power of Python to visualize, analyze, and model geological data. No experience in programming is required to use this book. If you are working in the Earth Sciences, are a novice programmer, and would like to exploit the power of Python in your projects, this is the right place for you.

Assisi, Italy   Maurizio Petrelli
March 2021

Acknowledgments

I would like to acknowledge all the people who encouraged me when I started planning this project and all those who supported me during the writing. The first one is Diego Perugini, who allowed me to re-enter academia in 2014 through the Chronos project after a hiatus of three years. I also thank the Erasmus Plus (E+) program that supported my foreign teaching excursions in Hungary and Germany, including Roberto Rettori and Sabrina Nazzareni, who oversaw the E+ program for my department, and the local E+ officers at the University of Perugia (Sonia Trinari and Francesca Buco) and at the Tiber Umbria Comett Education Programme (Maria Grazia Valocchia). Professor Francois Holtz (Leibniz Universität Hannover) and Professor Szabolcs Harangi (Eötvös University Budapest) are also kindly acknowledged for allowing me to run the "Python in Earth Sciences" courses at their institutions. The Department of Physics and Geology at University of Perugia, who supported this book through the Engage FRB2019 project, also has my gratitude. I also give my heartfelt thanks to my family, who put up with me as I wrote this book. Finally, I sincerely thank Aviva Loew (Academic Language Experts), Giuseppe la Spina, Eleonora Carocci, and Diego González-García for their critical suggestions, which have largely improved this book.

Just before I starting writing these acknowledgments, I received a message on my smartphone stating that I had an appointment for a COVID-19 vaccination the following day (March 3, 2021), exactly a year to the day after the initial lockdown in Italy. I interpreted this notification as a message of belief. I hope that the worldwide vaccination campaigns that have been launched will signal the beginning of a new era of beloved "normality" in our lives and that the time for resilience against COVID-19 is coming to an end. Now should be a time of empathy, cooperation, and rebirth.

Overview

Let me Introduce Myself

Hi and welcome. My name is Maurizio Petrelli and I currently work at the Department of Physics and Geology, University of Perugia (UniPg). My research focuses on the petrological characterization of volcanoes with an emphasis on the dynamics and timescales of pre-eruptive events. For this work, I combine classical and unconventional techniques. Since 2002, I've worked intensely in the laboratory, mainly focusing on the development UniPg's facility for Laser Ablation Inductively Coupled Plasma Mass Spectrometry (LA-ICP-MS). In February 2006, I obtained my Ph.D. degree with a thesis entitled "Nonlinear Dynamics in Magma Interaction Processes and their Implications on Magma Hybridization." Currently, I am developing a new line of research at UniPg, Department of Physics and Geology, for applying Machine Learning techniques in Geology. Finally, I also manage the LA-ICP-MS laboratory at UniPg.

Organization of Book

The book is organized into five parts plus three appendixes. The Part I, entitled "Python for Geologists: A Kickoff," focuses on the very basics of Python programming, from setting up an environment for scientific computing to solving your first geology problems using Python. The Part II is entitled "Describing Geological Data" and explains how to start visualizing (i.e., making plots) and generating descriptive statistics, both univariate and bivariate. The Part III, entitled "Integrals and Differential Equations in Geology," discusses integrals and differential equations while highlighting various applications in geology. The Part IV deals with "Probability Density Functions and Error Analysis" applied to the evaluation and modeling of Earth Science data. Finally, the Part V, entitled "Robust Statistics and Machine Learning" analyzes data sets that depart from normality (statistically speaking) and the application of machine learning techniques to data modeling in the Earth Sciences.

Styling Conventions

I use conventions throughout this book to identify different types of information. For example, Python statements, commands, and variables used within the main body of the text are set in italics.

Consider the following quoted text as an example "There are many options to create multiple subplots in matplotlib. In my opinion, the easiest approach is to create an empty figure [i.e., *fig = plt.figure()*, then add multiple axes (i.e., subplots) by using the method *fig.add_subplot(nrows, ncols, index)*]. The parameters *nrows*, *ncols*, and *index* indicate the numbers of rows and columns (*ncols*) and the positional index. In detail, *index* starts at 1 in the upper-left corner and increases to the right. To better understand, consider the code listing 4.4."

A block of Python code is highlighted as follows:

```
1  import pandas as pd
2
3  #Example 1
4  my_dataset1 = pd.read_excel('Smith_glass_post_NYT_data.xlsx',
5                              sheet_name='Supp_traces')
```

Listing 1 Example of code listing in Python

Shared Codes

All code presented in this book is tested on the Anaconda Individual Edition ver. 2021.5 (Python 3.8.8) and is available at my GitHub repository (◯ Petrelli-m): ⚭ http://bit.ly/python_earth_science

Involvement and Collaborations

I am always open to new collaborations worldwide. Feel free to contact me by mail to discuss new ideas or propose a collaboration. You can also reach me through my personal website or by Twitter. I love sharing the content of this book in short courses everywhere. If you are interested, please contact me to organize a visit to your institution.

Personal contacts:
✉ maurizio.petrelli@unipg.it
🐦 @mauripetre
↗ https://www.mauriziopetrelli.info

Contents

Part V   Robust Statistics and Machine Learning

Part I
Python for Geologists: A Kickoff

Chapter 1
Setting Up Your Python Environment, Easily

1.1 The Python Programming Language

Python is a high-level, modular, interpreted programming language.[1] What does this mean? A high-level programming language is characterized by a strong abstraction that cloaks the details of the computer so that the code is easy to understand for humans. Python is modular, which means that it supports modules and packages that allow program flexibility and code reuse. In detail, Python is composed of a "core" that deals with all basic operations plus a wide ecosystem of specialized packages to perform specific tasks. To be clear, a Python package or library is a reusable portion of code, which is a collection of functions and modules (i.e., a group of functions) allowing the user to complete specialized tasks such as reading an excel file or drawing a publication-ready diagram.

Python is an interpreted language (like MATLAB, Mathematica, Maple, and R). Conversely, C or FORTRAN are compiled languages. What is the difference between compiled and interpreted languages? Roughly speaking, with compiled languages, a translator compiles each code listing in an executable file. Once compiled, any target machine can directly run the executable file. Interpreted languages compile code in real time during each execution. The main difference for a novice programmer is that interpreted code typically runs slower than compiled executable code. However, performance is not an issue in most everyday operations. Performance starts becoming significant in computing-intensive tasks such as complex fluid dynamic simulations or three-dimensional (3D) graphical applications. If needed, the performance of Python can be significantly improved with the support of specific packages such as Numba, which can compile Python code. In this case, Python code approaches the speed of C and FORTRAN.

Being an interpreted language, Python facilitates code exchange over different platforms (i.e., cross-platform code exchange), fast prototyping, and great flexibility.

---

[1] https://www.python.org.

M. Petrelli, *Introduction to Python in Earth Science Data Analysis*,
Springer Textbooks in Earth Sciences, Geography and Environment,
https://doi.org/10.1007/978-3-030-78055-5_1

Some convincing arguments for Earth scientists to start learning Python are that (1) its syntax is easy to learn; (2) it is highly flexible; (3) it enjoys the support of a large community of users and developers; (4) it is free and open-source; and (5) it will improve your skills and proficiency.

1.2 Programming Paradigms

A programming paradigm is a style or general approach to writing code [7, 18, 19]. As a zeroth-order approximation, two archetypal paradigms dominate programming: imperative and declarative. Imperative programming mainly focuses on "how" to solve a problem, whereas declarative programming focuses on "what" to solve. Starting from these two archetypes, programmers have developed many derived paradigms, such as procedural, object-oriented, functional, logic, or aspect-oriented, just to to cite a few. The selection of a specific programming paradigm to develop your code depends on the overall nature of your project and final scope of your work. For parallel computing, the functional approach provides a well-established framework. However, given that an exhaustive documentation about programming paradigms is beyond the scope of this book, I will only illustrate those paradigms that are supported in Python.

The Python programming language is primarily designed for object-oriented programming, although it also supports, sometime spuriously, purely imperative, procedural, and functional paradigms [7, 18, 19]:

Imperative. The imperative approach is the oldest and simplest programming paradigm; one simply provides a defined sequence of instructions to a computer.

Procedural. The procedural approach is a subset of imperative programming. Instead of simply providing a sequence of instructions, it stores portions of code in one or more procedures (i.e., subroutines or functions). Any given procedure can be called at any point during the program execution, allowing for code organization and reuse.

Object-oriented. Like the procedural style, the object-oriented approach is a subset (i.e., an evolution) of imperative programming. Objects are the key elements in object-oriented programming. One of the main benefits of this approach is that it maintains a strong relation with real-world entities (e.g., shopping carts in websites, WYSIWYG environments).

Functional. The functional approach is a declarative type of programming. The purely functional paradigm bases the computation on evaluating mathematical functions and is well suited for high-load, parallel computing applications.

In this introductory book we will take advantage of Python's flexibility without focusing too much on specific code styling or on a particular paradigm. Specifically, our code remains mainly imperative for the easiest tasks but becomes more procedural for more advanced modeling. Also, we benefit from the many object-oriented libraries (e.g., pandas and matplotlib) developed for Python.

1.3   A Local Python Environment for Scientific Computing

Two main strategies are available to create a Python environment suitable for scientific computing on your personal computer: (1) install the Python core and add all required scientific packages separately; or (2) install a "ready-to-use" Python environment, specifically developed for scientific purposes. You can try both options but I suggest starting with option (2) because it requires almost zero programming skills and you will be ready to immediately and painlessly start your journey in the world of Python.

An example of a "ready-to-use" scientific Python environment is the Anaconda Python Distribution.[2] Anaconda Inc. (previously Continuum Analytics) develops and maintains the Anaconda Python distribution, providing different solutions that include a free release and two pay versions. The Individual Edition is the free option (and our choice); it is easy to install and offers community-driven support. **To install the Individual Edition of the Anaconda Python distribution, I suggest following the directives given in the official documentation.**[3] First, download and run the most recent stable installer for your Operating System (i.e., Windows OS, Mac OS or Linux). For Windows and Mac OS, a graphical installer is available. The installation procedure is the same as for any other software application. The Anaconda installer automatically installs the Python core and Anaconda Navigator, plus about 250 packages defining a complete environment for scientific visualization, analysis, and modeling. Over 7500 additional packages can be installed individually, as the need arises, from the Anaconda repository with the "conda"[4] package management system.

The Anaconda Navigator is a desktop graphical user interface (GUI), which means that it is a program that allows you to launch applications, install packages, and manage environments without using command-line instructions (Fig. 1.1).

From the Anaconda Navigator, we can launch two of the main applications that we will use to write code, run the modeling, and visualize the results. They are the Spyder application and the JupyterLab.

Spyder[5] is an Integrated Development Environment (IDE), i.e., a software application, providing a set of comprehensive facilities for software development and scientific programming. It combines a text editor to write code, inspection tools for debugging, and an interactive Python console for code execution. We will spend most of our time using Spyder. Figure 1.2 shows a screenshot of the Spider IDE.

JupyterLab is a web-based development environment to manage Jupyter Notebooks, which are web applications that allow you to create and share documents containing live code, equations, visualizations, and narrative text. Figure 1.3 shows a screenshot of a Jupyter Notebook.

---

[2] https://www.anaconda.com.

[3] https://www.anaconda.com/products/individual/.

[4] https://docs.conda.io/.

[5] https://www.spyder-ide.org.

Fig. 1.1 Screenshot of the Anaconda Navigator

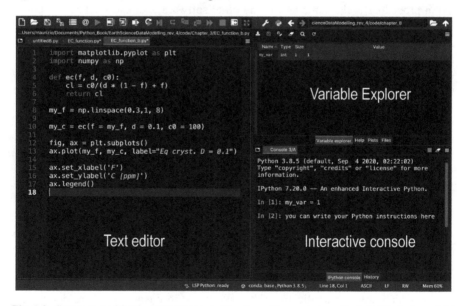

Fig. 1.2 Screenshot of Spider IDE. The text editor for writing code is on the left. The bottom-right panel is the IPython interactive console, and the top-right panel is the Variable Explorer

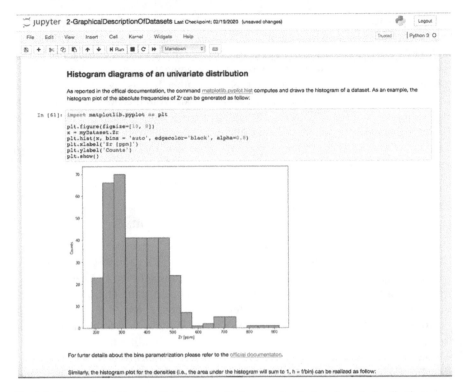

Fig. 1.3 Screenshot of Jupyter Notebook combining narrative text, code, and visualizations

Both Sypder and JupyterLab allow you to write code, perform computations, and report the results. There is not a preferred choice. My personal choice is to use Spyder and Jupyter Lab for research and teaching, respectively.

1.4 Remote Python Environments

Remote Python environments are those running in a computer system or virtual machines that can be accessed online. As an example, the Python environment can be installed on remote machines hosted by your academic institution (most universities have computing centers that offer this opportunity) or by commercial providers (often offering a basic free plan). The concepts and procedures described above for installing a local Python environment remain valid for remote machines. However, working with remote machines requires additional skills to access and operate online (e.g., knowledge of Secure Shell or Remote Desktop protocols for Linux and Windows-based machines, respectively). Therefore, to keep things simple, I suggest again starting with a local installation of the Anaconda Python distribution.

An alternative possibility to start working with Python online without installing a local environment is to use a remote IDE. For example, commercial providers such as Repl.it[6] and PythonAnywhere[7] offer free and complete Python IDEs, allowing the user to start coding first and later move on to developing more advanced applications. A drawback of this approach, however, is that neither IDE is specifically designed for scientific computing. Consequently, running code from this book will require the installation of additional libraries not included by default in the core distribution. Therefore, to easily replicate the code and examples given in this book, I suggest, once again, to locally install the most recent Anaconda Python distribution on your computer.

1.5 Python Packages for Scientific Applications

A key feature of Python is its modular nature. This section lists a few general-purpose scientific packages that we will make wide use of in this book. For each library, I provide with a quick description taken from the official documentation, a link to the official website and, when possible, a reference for further reading.

NumPy is a Python library that provides a multidimensional array object and an assortment of routines for fast operations on arrays, including mathematical, logical, shape manipulation, sorting, selecting, input-output, discrete Fourier transforms, basic linear algebra, basic statistical operations, random simulations, and other functionalities [3].[8]

Pandas is an open-source library providing high-performance, easy-to-use data structures and data-analysis tools for the Python programming language [4].[9]

SciPy is a collection of mathematical algorithms and functions built on the NumPy extension of Python. It adds significant power to interactive Python sessions by providing the user with high-level commands and classes for manipulating and visualizing data. With SciPy, an interactive Python session becomes a data-processing and system-prototyping environment rivaling systems such as MATLAB, IDL, Octave, R, and SciLab [3].[10]

Matplotlib is a Python library for creating static, animated, and interactive data visualizations [2].[11]

SymPy is a Python library for symbolic mathematics. Symbolic computation deals with the symbolic computation of mathematical objects. This means that mathemati-

---

[6] https://repl.it.

[7] https://www.pythonanywhere.com.

[8] https://numpy.org.

[9] https://pandas.pydata.org.

[10] https://scipy.org.

[11] https://matplotlib.org.

cal objects are represented exactly, not approximately, and mathematical expressions with unevaluated variables are left in symbolic form [12].[12]

Scikit-learn is an open-source machine learning library that supports supervised and unsupervised learning. It also provides various tools for model fitting, data pre-processing, model selection and evaluation, and many other utilities [13].[13]

1.6   Python Packages Specifically Developed for Geologists

Many Python packages have been developed to solve geology problems. They form a wide, heterogeneous, and useful ecosystem allowing us to achieve specific geology tasks. Examples include Devito, ObsPy, and Pyrolyte, to cite a few. Most of these packages can be easily installed by using the Conda package management system. Others requires a few additional steps and skills. The use of these specific packages is not covered in the present book, since they are typically developed to solve very specific geology problems. However, a novice to Python will benefit and probably require the notions reported in this book to be able to use these packages. Appendix A and the online repository[14] of the book provide a comprehensive list of resources and Python packages developed to solve geology tasks.

---

[12] https://www.sympy.org.

[13] https://scikit-learn.org.

[14] https://github.com/petrelli-m/python_earth_science_book.

Chapter 2
Python Essentials for a Geologist

2.1 Start Working with IPython Console

The IPython Console (Fig. 2.1) allows us to execute single instructions, multiple lines of code, and scripts, all of which may receive output from Python [15].

To start working with the IPython Console, consider Fig. 2.2, where the first two instructions are $A = 1$ and $B = 2.5$. The meaning of these two commands is straightforward: they simply assign the value of 1 and 2.5 to the variables A and B, respectively. The third instruction is $A + B$, which sums the two variables A and B, obtaining the result 3.5.

Figure 2.2 also provides information about the type of variables in Python (Fig. 2.3). For numbers, Python supports integers, floating point, and complex numbers. Integers and floating-point numbers differ by the presence or absence of decimals. In our case, A is an integer and B is a floating-point number. Complex numbers have a real part and an imaginary part, and they are not discussed in this book. Operations like addition or subtraction automatically convert integers into floating-point numbers if one of the operands (in our case, B) is floating point. The *type()* function returns the type of a variable. Additional data types that are relevant for this book are (a) Boolean (i.e., True or False), (b) Sequences, and (c) Dictionaries.

In Python, a Sequence type is an ordered collection of elements. Examples of sequences are Strings, Lists, and Tuples. Strings are sequences of characters, Lists are ordered collections of data, and Tuples are similar to Lists, but they cannot be modified after their creation. Figure 2.4 shows how to define and access Strings, Lists, and Tuples.

The elements of a sequence can be accessed by using indexes. In Python, the first index of a sequence is always 0 (zero). For example, the instruction *my_string*[0] returns the first element of *my_string* defined in Fig. 2.4 (i.e., "M"). Similarly, *my_touple*[2] returns the third element of *my_touple* (i.e., "Maurizio"). Additional examples on how to access a sequence are reported in Fig. 2.5. Using negative numbers (e.g., *my_string*[−1]), the indexing of the sequences starts from the last element and proceeds in reverse mode. Two numbers separated by a colon (e.g. [3:7]) define an index range, sampling the sequence from the lower to the upper bounds, excluding the upper bound. For the statement *my_string*[3 : 7], the interpreter samples

M. Petrelli, *Introduction to Python in Earth Science Data Analysis*,
Springer Textbooks in Earth Sciences, Geography and Environment,
https://doi.org/10.1007/978-3-030-78055-5_2

Fig. 2.1 IPython console

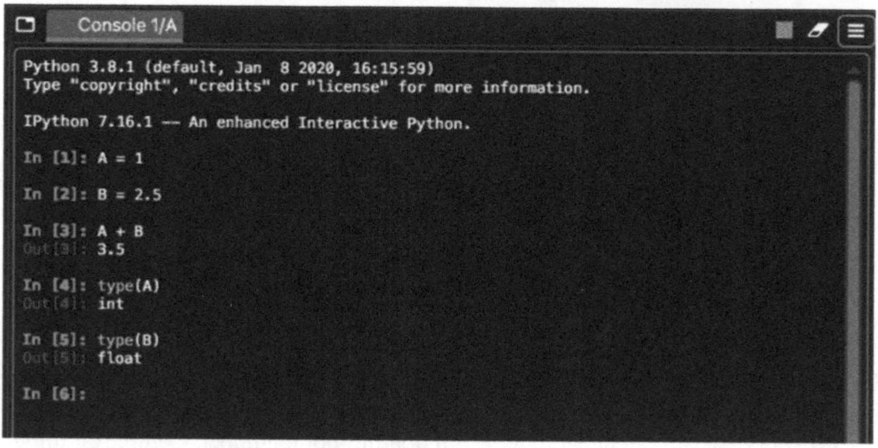

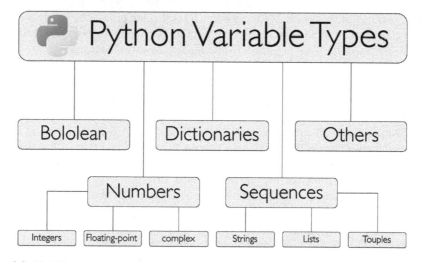

Fig. 2.2 Start working with the IPython console

Fig. 2.3 Variable data types in Python

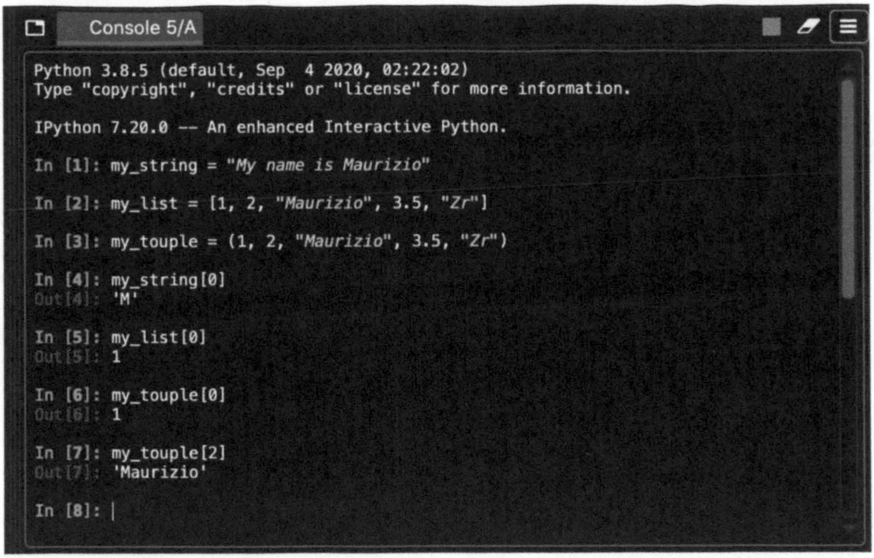

```
┌─┐    Console 5/A                                        ■  ✎  ≡

Python 3.8.5 (default, Sep  4 2020, 02:22:02)
Type "copyright", "credits" or "license" for more information.

IPython 7.20.0 -- An enhanced Interactive Python.

In [1]: my_string = "My name is Maurizio"

In [2]: my_list = [1, 2, "Maurizio", 3.5, "Zr"]

In [3]: my_touple = (1, 2, "Maurizio", 3.5, "Zr")

In [4]: my_string[0]
Out[4]: 'M'

In [5]: my_list[0]
Out[5]: 1

In [6]: my_touple[0]
Out[6]: 1

In [7]: my_touple[2]
Out[7]: 'Maurizio'

In [8]: |
```

Fig. 2.4 Defining and working with sequences

my_string from the third to the seventh indexes (i.e., "name"). Finally, commands like $my\_string[: 2]$ and $my\_string[11 :]$ sample my_string from the beginning to the index 2 (excluded) and from the index 11 to the last element, respectively.

Dictionaries are data types consisting of a collection of key-value pairs. A dictionary can be defined by enclosing a comma-separated list of key-value pairs in curly braces, with a colon separating each key from the associated value (Fig. 2.6). In a dictionary, a value is retrieved by specifying the corresponding key in square brackets (Fig. 2.6).

2.2 Naming and Style Conventions

The main aim of using conventions in programming is to improve the readability of codes and to facilitate collaboration between different programmers. In Python, the "PEP 8—Style Guide for Python Code" gives the coding conventions for Python.[1]

Writing readable code is important for many reasons, the main one of which is to allow others to easily understand your code. This is crucial when working on collaborative projects. By sharing best practices, programming teams will write consistent and elegant codes.

In the book, I try to follow the main rules defined by the PEP 8—Style Guide for Python Code. Beginners should keep in mind that these guidelines exist and start

---

[1] https://www.python.org/dev/peps/pep-0008/.

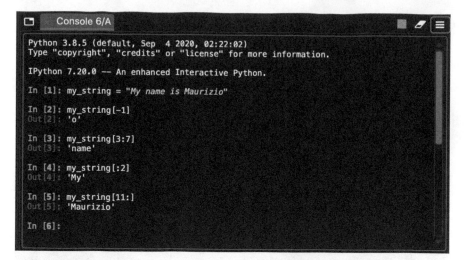

Fig. 2.5 Accessing sequences

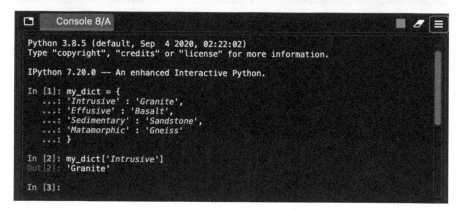

Fig. 2.6 Defining and accessing dictionaries

following the most important ones, which are listed in Table 2.1 for convenience. However, I suggest that beginners focus more on their results (i.e., achieving the objectives) than on the form of the code.

2.3 Working with Python Scripts

A script is a sequence of code instructions used to automate processes (e.g., making a diagram, or a geological model) that would otherwise need to be executed step by step (e.g., in the IPython console). In detail, Python scripts are text files typically

Table 2.1 Styling and Naming conventions in Python

Type	Style or Naming convention	Example
Function	Function names should be lowercase, with words separated by underscores as necessary to improve readability (cf. Sect. 2.4)	Function, my_function
Variable	Variable names follow the same convention as function names	x, my_dataset
Constant	Constants are usually written in all capital letters with underscores separating words	A, GREEK_P
Class	Start each word with a capital letter (CapWords convention). Do not use underscores to separate subsequent words (cf. appendix B)	Circle, MyClass
Method	Use the function naming rules: lowercase with words separated by underscores as necessary to improve readability (cf. appendix B)	Method, my_method
Names to avoid	Never use the characters 'l' (lowercase letter el), 'O' (uppercase letter oh), or 'I' (uppercase letter eye) as single character variable names	–
Indentation	PEP 8 recommends using four spaces per indentation level (cf. Sect. 2.4)	–

characterized by a .py extension and containing a sequence of Python instructions. Writing and modifying Python scripts requires nothing more than a text editor. Spyder incorporates a text editor with advanced features such as code completion and syntax inspection. In Spyder, the text editor is usually positioned in a panel on the left portion of the screen. To execute a Python script, the interpreter reads each instruction sequentially, starting from the first line. To execute a Python script in the active IPython console of Spyder, we click the play button, as shown in Fig. 2.7, or use the F5 keyboard shortcut. Keyboard shortcuts help us be more proficient; Table 2.2 lists a few additional keyboard shortcuts.

The script Listing 2.1 gives the Python script of Fig. 2.7 and, in lines 5 to 10, the output obtained upon running the script in the IPython console.

The three single quotation marks (i.e., ''') in lines 5 and 10 of the script Listing 2.1 open and close a multi-line comment, which is simply lines of code or text that are ignored by the interpreter. The symbol # means that the remaining text on the same line is a comment. Comments are a fundamental part of Python codes because they help you and future users clarify the code workflow. Keep in mind that you might spend an entire day developing a very proficient script only to wake up the next morning without remembering how the script works! Comments are a godsend in these situations.

In fact, you don't necessarily need Spyder to write a .py script. As stated above, Python scripts can be written using any text editor. The *python* instruction will run your scripts in the command line or terminal application (see Fig. 2.8).

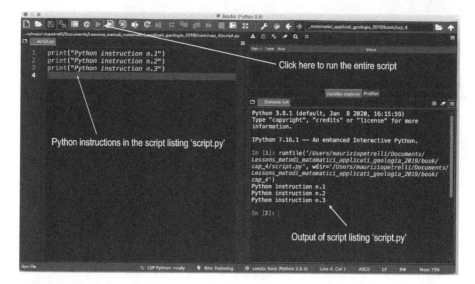

Fig. 2.7 Running a Python script in Spyder

Table 2.2 Selected spyder keyboard shortcuts

Windows OS	Mac OS	Action
F5	F5	Run file (complete script)
F9	F9	Run selection (or current line)
Ctrl + T	Cmd + T	Open an IPython console
Ctrl + space	Ctrl + space	Code completion
Tab	Tab	Indent selected line(s)
Shift + Tab	Shift + Tab	Unindent selected line(s)
Ctrl + Q	Cmd + Q	Quit Spyder

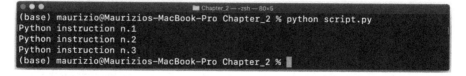

Fig. 2.8 Running a Python script using the *python* instruction in a MacBook Terminal application

```
1   print("Python instruction n.1")
2   print("Python instruction n.2")
3   print("Python instruction n.3")
4
5   '''
6   Output:
7   Python instruction n.1
8   Python instruction n.2
9   Python instruction n.3
10  '''
```

Listing 2.1 A simple script in Python

2.4 Conditional Statements, Indentation, Loops, and Functions

Conditional Statements

In Python, the *if* statement indicates the conditional execution of single or multiple instructions based on the value of an expression.

```
1   my_var = 2
2
3   if my_var > 2:
4       print('my_var is greater than 2')
5   elif my_var == 2:
6       print('my_var is equal to 2')
7       # more instructions could be added
8       # using the same indentation
9   else:
10      print('MyVar is less than 2')
11
12  '''
13  Output:
14  my_var is equal to 2
15  '''
```

Listing 2.2 If, elif, else statements

To understand, consider the script Listing 2.2. At line 1, we define the variable *my_var* and assign it the value 2. At line 3, the if statement evaluates *my_var* and executes the instruction at line 4 only if *my_var* is greater than 2. Given that this is not in this case, the interpreter jumps to line 5 and evaluates whether *my_var* is equal to 2. Note that "=" assigns a value to a variable, whereas "==" compares two quantities, returning "True" if they are equal and "False" if they differ. Given that *my_var* equals 2, the interpreter executes the instructions from line 6 to 8. Finally, the instruction at line 10 is executed in all remaining cases (i.e., when *my_var* is less than 2).

Indentation and Blocks

The term "indentation" refers to adding one or more white spaces before an instruction. In a Python script, contiguous instructions (e.g., lines 6 to 8 of the script Listing 2.2) that are indented to the same level are considered to be part of the same block of code. A code block is considered by the interpreter as a single entity, which allows us to structure Python scripts. For example, the blocks after the *if*, *elif*, and *else* statements in script Listing 2.2 are executed in accordance with the conditions given on lines 3, 5, and 9, respectively.

To better understand how indentation works in Python, consider the code Listing 2.3. The instructions from line 1 to 3 and at line 12 are always executed each time we run the script. The interpreter executes the instructions at lines 5, 9, 10, and 11 if and only if the variable a equals 1. Finally, the interpreter executes lines 7 and 8 if and only if a and b equal 1 and 3, respectively.

Note that indentation is a fundamental concept in Python, allowing us to define simple operations like conditional statements, loops, and functions, but also more complex structures like modules and packages.

```
1   # this instruction is always executed
2   # this instruction is always executed
3   # this instruction is always executed
4   if a == 1:
5       # this instruction is executed if a = 1
6       if b == 3:
7           # this instruction is executed if a = 1 and b = 3
8           # this instruction is executed if a = 1 and b = 3
9       # this instruction is executed if a = 1
10      # this instruction is executed if a = 1
11      # this instruction is executed if a = 1
12  # this instruction is always executed
```

Listing 2.3 Python uses indentation to define blocks of code

For Loops

The *for* loop in Python iterates over a sequence (i.e., lists, tuples, and strings) or other iterable objects. As an example, the code Listing 2.4 iterates over the list named *rocks*. At line 1 we define a list (i.e., *rocks*), at line 3 we implement the iteration, and at line 4 we print to the screen the result of each iteration, namely, each element of the sequence.

Often, we perform iterations using *range()*. The command *range()* is a Python function that returns a sequence of integers.

The range syntax is *range(start, stop, step)* where the arguments start, stop, and step are the initial, final, and step values of the sequence, respectively. Note that the upper limit (i.e., *stop*) is not included in the sequence. If we pass only one argument to the range function [e.g., *range(6)*], it is interpreted as the stop parameter, with the sequence starting from 0. The code Listing 2.5 shows some examples of iterations over sequences of numbers generated using the *range()* function.

```
1  rocks = ['sedimentary', 'igneous intrusive', 'igneous
       effusive', 'methamorphic']
2
3  for rock in rocks:
4      print(rock)
5
6  '''
7  Output:
8  sedimentary
9  igneous intrusive
10 igneous effusive
11 methamorphic
12 '''
```
Listing 2.4 Iterate over a list

```
1  print('a sequence from 0 to 2')
2  for i in range(3):
3      print(i)
4
5  print('--------------------')
6  print('a sequence from 2 to 4')
7  for i in range(2, 5):
8      print(i)
9
10 print('--------------------')
11 print('a sequence from 2 to 8 with a step of 2')
12 for i in range(2, 9, 2):
13     print(i)
14
15 '''
16 Output:
17 a sequence from 0 to 2
18 0
19 1
20 2
21 --------------------
22 a sequence from 2 to 4
23 2
24 3
25 4
26 --------------------
27 a sequence from 2 to 8 with a step of 2
28 2
29 4
30 6
31 8
32 '''
```
Listing 2.5 Iterating over a sequence of numbers generated using *range()*

While Loops

The *while* loop begins by checking a test condition and starts only if the test-condition is True. After each iteration over the loop instructions, the test condition is checked again and the loop continues until the test-condition is False. To better understand, consider the code Listing 2.6. At line 1, we define the object *my_var* and assign it the value 0. At line 3, we evaluate the test condition *my_var* < 5. Given that *my_var* equals 0, the test condition is True and the interpreter enters the loop. At line 4, it prints *my_var* (i.e., 0), and at line 5, *my_var* is assigned the value 1. The loop then returns to line 3, where the test condition is evaluated again, and continues as long as the test condition remains True (i.e., as long as *my_var* < 5). Consequently, the interpreter repeats the instructions at lines 4 and 5 (i.e., the block of code with the same indentation after the test condition) until *my_var* is assigned the value 5.

```
1   my_var = 0
2
3   while my_var < 5:
4       print(my_var)
5       my_var = my_var + 1
6
7   '''
8   Output:
9   0
10  1
11  2
12  3
13  4
14  '''
```

Listing 2.6 While loops

Functions

A function is a block of reusable code that is developed to complete a specific task. Function blocks begin with the keyword *def* followed by the name of the function and then parentheses (see code Listing 2.7). Input parameters or arguments should be placed within these parentheses. The code block of a function starts after a colon (:) and must be indented. By using the optional statement *return*, we can pass back to the caller a single or multiple answers, such as some variables computed within the function. The code Listing 2.7 shows how to define and use a simple function. At line 1, we define a function named *sum* that accepts the two arguments a and b. At line 2, the function assigns the sum of a and b to the variable c. Finally, the function ends at line 3, returning c to the caller. At line 5, we define the variable res by calling the function sum with a = 2 and b = 3 as arguments. At line 6, we print a string containing the value of res [note that the *str()* function converts numbers to strings].

```
1  def sum(a, b):
2      c = a + b
3      return c
4
5  res = sum(a=2, b=3)
6  print('the result is ' + str(res))
7
8  '''
9  Output:
10 the result is 5
11 '''
```

Listing 2.7 Defining a function

2.5 Importing External Libraries

The Anaconda Python distribution includes almost all the packages required for the most common operations in Data Science, such as NumPy, SciPy, pandas, matplotlib, seaborn, and scikit-learn, which are briefly described in Sect. 1.5. The *import* and *from* statements allow us to import entire modules, packages, or single functions into our scripts. The code Listing 2.8 gives examples of using the *import* and *from* statements. Note that, at line 1 of the code Listing 2.8, the entire pandas package is imported into an object named *pd*. At line 2, we import the matplotlib package pyplot into the object plt. At line 3, the *random()* function is imported from the NumPy package. Finally, at line 4, we import all the functions in the SymPy package by using the wildcard character "*" (note that the use of the wildcard character "*" for the importing is currently discouraged since it does not specify which items are imported, possibly leading to problems, especially in large projects).

```
1  import pandas as pd
2  import matplotlib.pyplot as plt
3  from numpy import random
4  from sympy import * # To note: import * should be avoided
```

Listing 2.8 Use of *import* and *from* statements

2.6 Basic Operations and Mathematical Functions

Basic mathematical operators such as sum or multiplication are always available in Python and are listed in Table 2.3. Additional trigonometric and arithmetic functions are available by importing the math and NumPy libraries, which also contain relevant constants such as π (Archimedes' constant) and e (Euler's number). The main difference between the math and NumPy libraries is that the former is designed to work with scalars whereas the latter is designed to work with arrays. However, NumPy works well with scalars, too. Given that NumPy is more flexible than math, the examples in this book use exclusively the NumPy library.

Table 2.3 Basic mathematical operations in Python

Operator	Description	Example	Operator	Description	Example
+	Addition	$3 + 2 = 5$	−	Subtraction	$3 - 2 = 1$
∗	Multiplication	$3 * 2 = 6$	/	Division	$6 / 2 = 3$
∗∗	Power	$3 ** 2 = 9$	%	Modulus	$2 \% 2 = 0$

```python
1   import numpy as np # import numpy
2
3   # relevant constants
4   GREEK_P = np.pi
5   EULER_NUMBER = np.e
6
7   # print greek_p and euler_number on the screen
8   print("Archimedes' constant is " + str(GREEK_P))
9   print("Euler's number is " + str(EULER_NUMBER))
10
11  # trigonometric functions
12  x = np.sin(GREEK_P / 2) # x = 1 expected
13
14  # print the result on the screen
15  print("The sine of a quarter of radiant is " + str(x))
16
17
18  # defining a 1D array in numpy
19  my_array = np.array([4, 8, 27])
20  # print myArray on the screen
21  print("myArray is equal to " + str(my_array))
22
23  log10_my_array = np.log10(my_array)
24
25  # print the result on the screen
26  print("The base-10 logarithms of the elements in myArray
          are")
27  print(log10_my_array)
28
29  '''
30  Output:
31  Archimedes' constant is 3.141592653589793
32  Euler's number is 2.718281828459045
33  The sine of a quarter of radiant is 1.0
34  myArray is equal to [ 4  8 27]
35  The base-10 logarithms of the elements in myArray are
36  [0.60205999 0.90308999 1.43136376]
37  '''
```

Listing 2.9 Our first exposure to NumPy

Tables 2.4 and 2.5 list some relevant NumPy constants and functions, respectively. Also, code Listing 2.9 provides some introductory examples that illustrate how to use NumPy constants and mathematical functions.

We are now ready to begin learning how to use Python to solve geology problems.

Table 2.4 Relevant constants in NumPy

NumPy	Description	Value	NumPy	Description	Value
e	Euler's number (e)	2.718...	pi	Archimedes' const. (π)	3.141...
euler_gamma	Euler's constant (γ)	0.577...	inf	Positive infinity	∞

Table 2.5 Introducing exponents, logarithms, and trigonometric functions in NumPy

NumPy	Description	NumPy	Description	NumPy	Description
sin()	Trigonom. sine	cos()	Trigonom. cosine	tan()	Trigonom. tangent
arcsin()	Inverse sine	arccos()	Inverse cosine	arctan()	Inverse tangent
exp()	Exponential	log()	Natural logarithm	log10()	Base 10 logarithm
log2()	Base-2 logarithm	sqrt()	Square-root	abs()	Absolute value

Chapter 3
Solving Geology Problems Using Python: An Introduction

3.1 My First Binary Diagram Using Python

To start learning Python we will analyze geological data by applying two basic operations: importing data sets using the pandas library and representing them in binary diagrams. As introduced in Sect. 1.5, pandas is a Python library (i.e., a tool) designed to facilitate working with structured data. In practice, it provides a host of ready-to-use functions to work with scientific data. For example, with a single line of code, we can use pandas to import a data set stored in an Excel spreadsheet or in a text file. To understand how this is done, consider the code Listing 3.1.

```
1  import pandas as pd
2
3  #Example 1
4  my_dataset1 = pd.read_excel('Smith_glass_post_NYT_data.xlsx',
5                              sheet_name='Supp_traces')
```

Listing 3.1 Importing data from an Excel file into Python

At line 1, we import the pandas library into an object named *pd*, which we use to store all of pandas' functionalities.

At line 4, we define a pandas DataFrame (i.e., *my_dataset1*) into which we read data from an Excel file named "Smith_glass_post_NYT_data.xlsx." Also, since the Excel file potentially contains several spreadsheets, we specify the spreadsheet "Supp_traces." The imported data set contains the chemical concentrations of trace elements of volcanic tephra published by [16] and will serve as a proxy for a geological data set. In detail, it consists of major (Supp_majors) and trace-element (Supp_traces) analyses of tephra samples belonging to the recent activity (last 15 ky) of the Campi Flergrei Caldera (Italy Fig. 3.1).

The instruction *pd.read_excel()* accepts numerous arguments, which gives it significant flexibility. The two most important are a valid *string-path* and the

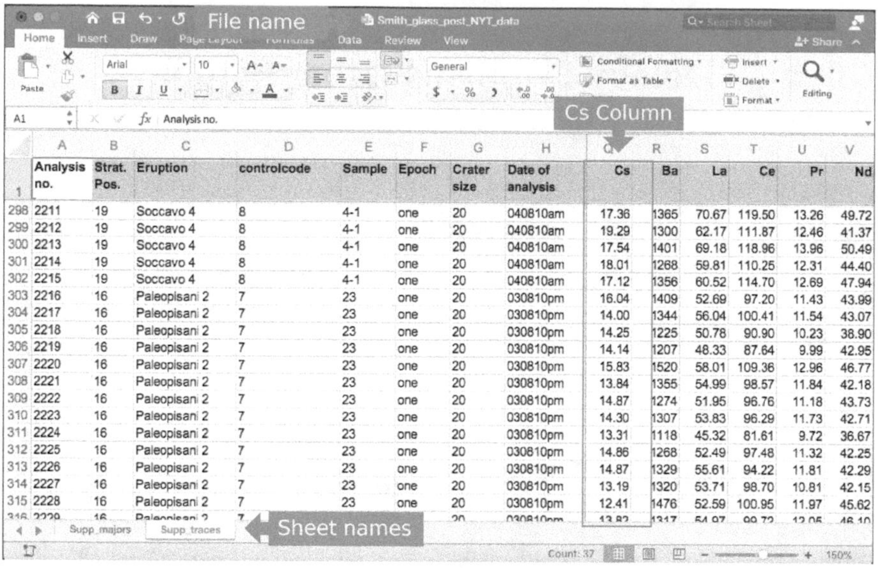

Fig. 3.1 The Smith_glass_post_NYT_data.xlsx Excel file

sheet_name. In our case, the *string-path* is the name of the Excel file (i.e., "Smith_glass_post_NYT_data.xlsx"). If we only provide the file name in *string-path*, the Excel file must be in the same folder as the Python script. Additional allowed values for *string-path* are local file addresses (e.g., "/Users/maurizio/Documents/file.xlsx") or valid URL schemes, including http, ftp, and s3. The *sheet_name* parameter can be a string, an integer, a list, or None. The default value is 0, meaning that *pd.read_excel()* opens the first sheet of the Excel file. In detail, integers and strings indicate sheet positions starting from 0 and sheet names, respectively. Finally, lists of strings or integers are used to request multiple sheets.

Return now to line 4 of the code Listing 3.1, where we defined a DataFrame. What is a DataFrame? It is "a two-dimensional labeled data structure with columns of potentially different types,"[1] which means that we can envision a DataFrame as a simple data table over which Python has full control.

To start plotting, we introduce an additional library named matplotlib, which is "a comprehensive library for creating static, animated, and interactive visualizations in Python. It is a Python two dimensional (2D) plotting library that produces publication-quality figures in a variety of hardcopy formats and interactive environments across platforms."[2] With just a few lines of code, it generates plots, histograms, power spectra, bar charts, scatter plots, etc. Matplotlib allows two different coding styles: pyplot and object-oriented Application Programming Interfaces (APIs). In matplotlib.pyplot (i.e., pyplot-style), each function changes the active figure. In

---

[1] https://pandas.pydata.org/pandas-docs/stable/user_guide/dsintro.html.

[2] https://matplotlib.org.

practice, each command produces an effect on your diagram, allowing it to be easily organized and managed in the imperative coding paradigm (i.e., the most basic and easiest coding paradigm; cf. Sect. 1.2). As a drawback, matplotlib.pyplot is less flexible and less powerful than the matplotlib object-oriented interface (i.e., the OO-style), which is not more difficult to learn than pyplot. Consequently, I suggest starting to familiarize yourself with the OO-style directly with easy examples, and then going in deeper detail (see Appendix C).

As an example, the code Listing 3.2 shows how to make a simple binary diagram using the OO-style API. In detail, code Listing 3.2 shows how to plot the elements Th versus Zr in a scatter diagram. The workflow is simple: at lines 1 and 2, we import the pandas library and matplotlib.pyplot module, respectively. As discussed above, line 4 imports the Excel file "Smith_glass_post_NYT_data.xlsx" into a DataFrame named *my_dataset*1. At lines 6 and 7, we define two data sequences by selecting the columns Zr and Th, respectively, from *my_dataset*1. At line 9, we generate a "figure" (i.e., the object *fig*) containing only one "axes" (*ax*). Note that, in matplotlib, the figure object represents the whole diagram whereas the "axes" are what you typically think of when using the word "plot" (see Appendix C). A given figure can host a single axis (i.e., a simple diagram) or many axes (i.e., a figure containing two or more sub-plots).

```
1   import pandas as pd
2   import matplotlib.pyplot as plt
3
4   my_dataset1 = pd.read_excel('Smith_glass_post_NYT_data.xlsx',
        sheet_name='Supp_traces')
5
6   x = my_dataset1.Zr
7   y = my_dataset1.Th
8
9   fig, ax = plt.subplots() # Create a figure containing one axes
10  ax.scatter(x, y)
```
Listing 3.2 Our first attempt to make a binary diagram in Python

Although the diagram reported in Fig. 3.2 is a good start for a novice, it is missing significant mandatory information (e.g., axis label). To add features to the diagram, we use *ax.set_title*(), *ax.set_xlabel*(), and *ax.set_ylabel*() to add a title and labels to the *x* and *y* axes, respectively (code Listing 3.3). Figure 3.3 shows the diagram of Fig. 3.2 with a new title and axis labels.

To improve our skills in the use of Python to visualize scientific data, consider at the code Listing 3.4, which shows how to slice a data set. In detail, lines 2 and 3 divide the original data set (i.e., my_dataset1) into two sub data sets (i.e., *my_sub_dataset*1 and *my_sub_dataset*2) characterized by Zr contents above (i.e., ">", line 2) and below (i.e., "<", line 3) 450 ppm, respectively. Note that if you would like to include Zr = 450 ppm to one of the two sub-datasets, you should use the operators ">=" and "<=", respectively.

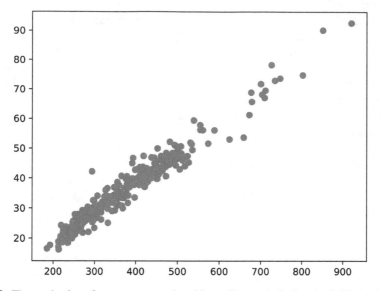

Fig. 3.2 The result of our first attempt to make a binary diagram in Python (code Listing 3.2)

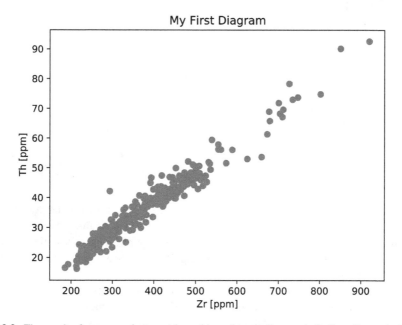

Fig. 3.3 The result of our second attempt in making a binary diagram in Python; it now includes a title and axis labels (code Listing 3.3)

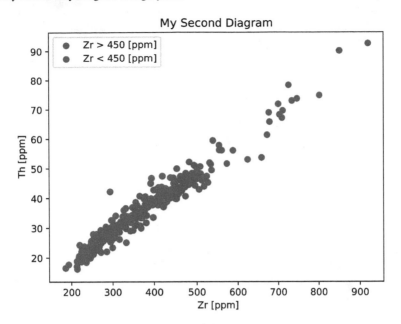

Fig. 3.4 The result of the code given in Listing 3.4

```
1  fig, ax = plt.subplots()
2  ax.scatter(x, y)
3  ax.set_title("My First Diagram")
4  ax.set_xlabel("Zr [ppm]")
5  ax.set_ylabel("Th [ppm]")
```

Listing 3.3 Our second attempt to make a binary diagram in Python, which includes a title and axis labels

The sub data sets *my_sub_dataset*1 and *my_sub_dataset*2 are then plotted at lines 11 and 16, respectively. Note that all the plotting instances (i.e., lines 11 and 16) that occur after the command *plt.subplots()* display the results in the same figure. Figure 3.4 shows the result of code Listing 3.4.

```
1  # Define two sub-dataset for Zr>450 and Zr<450 respectively
2  my_sub_dataset1 = my_dataset1[my_dataset1.Zr > 450]
3  my_sub_dataset2 = my_dataset1[my_dataset1.Zr < 450]
4
5  #generate a new picture
6  fig, ax = plt.subplots()
7  # Generate the scatter Zr Vs Th diagram for Zr > 450
8  # in blue also defining the legend caption as "Zr > 450 [ppm]"
9  x1 = my_sub_dataset1.Zr
10 y1 = my_sub_dataset1.Th
11 ax.scatter(x1, y1, color='blue', label="Zr > 450 [ppm]")
12 # Generate the scatter Zr Vs Th diagram for Zr < 450
13 # in red also defining the legend caption as "Zr < 450 [ppm]"
14 x2 = my_sub_dataset2.Zr
```

```
15  y2 = my_sub_dataset2.Th
16  ax.scatter(x2, y2, color='red', label="Zr < 450 [ppm]")
17
18  ax.set_title("My Second Diagram")
19  ax.set_xlabel("Zr [ppm]")
20  ax.set_ylabel("Th [ppm]")
21  # generate the legend
22  ax.legend()
```

Listing 3.4 Making a binary diagram with a sub-sampling (i.e., Zr > 450 and Zr < 450 ppb) of the original data set

```
1   fig, ax = plt.subplots()
2
3   my_data1 = my_dataset1[(my_dataset1.Epoch == 'one')]
4   ax.scatter(my_data1.Zr, my_data1.Th, label='Epoch 1')
5
6   my_data2 = my_dataset1[(my_dataset1.Epoch == 'two')]
7   ax.scatter(my_data2.Zr, my_data2.Th, label='Epoch 2')
8
9   my_data3 = my_dataset1[(my_dataset1.Epoch == 'three')]
10  ax.scatter(my_data3.Zr, my_data3.Th, label='Epoch 3')
11
12  my_data4 = my_dataset1[(my_dataset1.Epoch == 'three-b')]
13  ax.scatter(my_data4.Zr, my_data4.Th, label='Epoch 3b')
14
15  ax.set_title("My Third Diagram")
16  ax.set_xlabel("Zr [ppm]")
17  ax.set_ylabel("Th [ppm]")
18  ax.legend()
```

Listing 3.5 Binary diagram with a sub-sampling (i.e., using the labels of the Epoch column) of the original data set

```
1   epochs = ['one','two','three','three-b']
2
3   fig, ax = plt.subplots()
4   for epoch in epochs:
5       my_data = my_dataset1[(my_dataset1.Epoch == epoch)]
6       ax.scatter(my_data.Zr, my_data.Th, label="Epoch " + epoch)
7
8   ax.set_title("My Third Diagram again")
9   ax.set_xlabel("Zr [ppm]")
10  ax.set_ylabel("Th [ppm]")
11  ax.legend()
```

Listing 3.6 Re-writing the code of listing 3.5 using a *for* loop

We now continue with an additional example of DataFrame slicing. In detail, the code Listing 3.5 shows how to filter the original data set by using the labels given in the "Epoch" column.

These labels divide the eruptions in four different periods (i.e., one, two, three, and three-b). After the slicing the sub data sets (lines 3, 6, 9, and 12), samples belonging to different Epochs are plotted using unique labels (lines 4, 7, 10, and 13, respectively; Fig. 3.5).

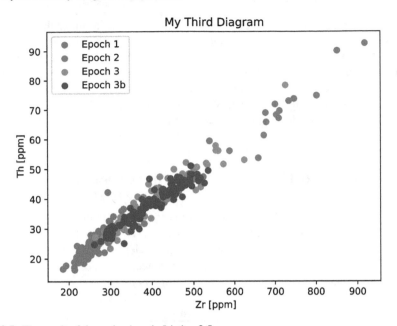

Fig. 3.5 The result of the code given in Listing 3.5

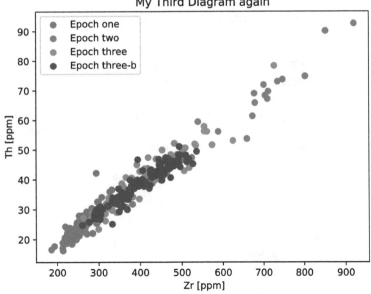

Fig. 3.6 The result of the code given in Listing 3.6

Readers already familiar with Python may suggest a way to make this code more concise and more elegant by using a loop (see code Listing 3.6 and Fig. 3.6).

As explained in Sect. 2.4, the *for* loop is used in Python to iterate over a sequence. Although you should become proficient in the use of loops, conditional statements, and functions (see Sect. 2.4), many everyday operations and tasks can be completed without a deep knowledge of the syntax and "core semantics" of the Python language. To see this, let's solve our first geology problem.

3.2 Making Our First Models in Earth Science

Developing a simple Earth Sciences model can provide useful information regarding Python syntax, workflow, and ideology. This section shows how to develop a simple function (Sect. 2.4) describing the evolution of trace elements in a magmatic system. In detail, Eq. (3.1) describes the evolution of the concentration (C) of a trace element in the liquid phase of a magmatic system during crystallization at thermodynamic equilibrium [14]:

$$C = \frac{C_0}{D(1 - F) + F}. \tag{3.1}$$

The quantities C_0, D, and F are the initial concentration, the bulk partition coefficient of the trace element between melt and crystals, and the relative amount of melt in the system, respectively. The code Listing 3.7 shows how to create a function to solve Eq. (3.1) for the concentration C.

In code Listing 3.7, we define at line 1 a function named *ec()* that accepts F, D, and C_0 as arguments. Note that the code at line 2 is indented (recall that "indented" refers to the spaces that appear at the beginning of a line of code; see Sect. 2.4). All subsequent lines in the function must have at a minimum the same indentation as this first line of code in the function. Thus, in the present case, lines 2 and 3 are both similarly indented and are thus part of the function *ec()*.

```
1   def ec(f, d, c0):
2       c1 = c0/(d * (1-f) + f)
3       return c1
4
5   my_c = ec(f=0.5, d=0.1, c0=100)
6
7   print('RESULT: '+ str(int(my_c)) + ' ppm')
8
9   '''
10  Output:
11  RESULT: 181 ppm
12  '''
```

Listing 3.7 Defining a function in Python to model Eq. (3.1)

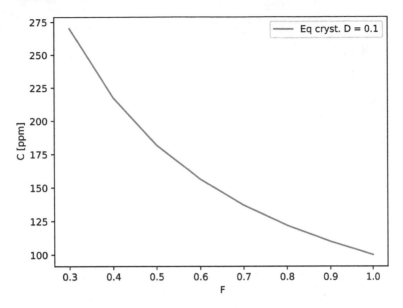

Fig. 3.7 The result of the code given in code Listing 3.8

The computation is done at line 2 and the result is returned at line 3. At line 5, we call the function *ec()*, which computes the concentration of a trace element in the melt phase of a system characterized by F, D, and C_0 equal to 0.5, 0.1 and 100 ppm, respectively. Line 7 prints the result, which appears at line 11 (181 ppm). Line 7 of code Listing 3.7 requires further explanation. The statement *print()* prints the content in the brackets to the screen, whereas the function *str()* converts a number to a string and the function *int()* truncates a decimal number to an integer.

Code Listing 3.8 presents a more in-depth investigation of Eq. (3.1) for F ranging from 1.0 to 0.3. For readers not familiar with Eq. (3.1), Fig. 3.7 shows the behavior of incompatible elements ($D < 1$, i.e., elements that do not easily enter crystals growing in the system but prefer to remain in the melt phase) in a completely molten system (i.e., $F = 1$) to a magmatic mush characterized by a relative amount of melt on the order of $F = 0.3$.

Line 1 of code Listing 3.8 is now straightforward: it imports the matplotlib.pyplot functionalities into the object *plt* for use in our script. At line 2, we import the NumPy library. As discussed in Sect. 1.5, NumPy is a package for scientific computing and can handle N-dimensional arrays, linear algebra, Fourier transform, random numbers, and other mathematical niceties. The meaning of lines 4–6 is also straightforward: they define the *ec()* function, as done in code Listing 3.7.

The use of NumPy starts at line 8 with the statement *np.linspace(0.3, 1, 8)*, which generates a one-dimensional (1D) array made of eight elements starting at 0.3 and ending at 1.0. Line 7 of code Listing 3.9 shows the result of printing *my_f* to the screen.

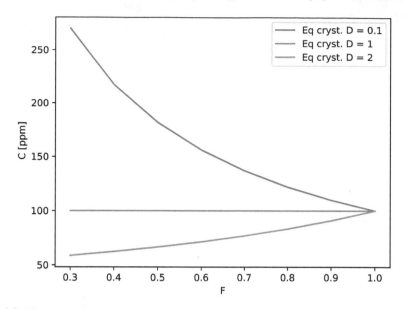

Fig. 3.8 The result of the code given in code Listings 3.11 and 3.12

Moving back to code Listing 3.8, we call at line 10 the *ec()* function with the arguments *my_f* (i.e., a 1D array of eight elements), 0.1, and 100 ppm for F, D, and C_0, respectively. The result is stored in *my_c* and is a 1D array of eight elements, one for each element of the array *my_f*. At line 13, we plot the results *my_f* versus *my_c* by using *ax.plot()*. By default, it plots a binary diagram connecting successive points by lines. Figure 3.7 shows the result of code Listing 3.8.

```
1   import matplotlib.pyplot as plt
2   import numpy as np
3
4   def ec(f, d, c0):
5       c1 = c0/(d * (1-f) + f)
6       return c1
7
8   my_f = np.linspace(0.3, 1, 8)
9
10  my_c = ec(f=my_f, d=0.1, c0=100)
11
12  fig, ax = plt.subplots()
13  ax.plot(my_f, my_c, label="Eq cryst. D = 0.1")
14
15  ax.set_xlabel('F')
16  ax.set_ylabel('C [ppm]')
17  ax.legend()
```

Listing 3.8 Solving Eq. (3.1) with F ranging from 0.3 to 1 and plotting the results

```
1  my_f = np.linspace(0.3, 1, 8)
2
3  print(my_f)
4
5  '''
6  Output:
7  [0.3 0.4 0.5 0.6 0.7 0.8 0.9 1. ]
8  '''
```

Listing 3.9 The np.linspace() statement

```
1  my_f = np.arange(0, 10, 1)
2
3  print(my_f)
4
5  '''
6  Output:
7  [0 1 2 3 4 5 6 7 8 9]
8  '''
```

Listing 3.10 The np.arange() statement

```
1   import matplotlib.pyplot as plt
2   import numpy as np
3
4   def ec(f, d, c0):
5       c1 = c0/(d * (1-f) + f)
6       return c1
7
8   my_f = np.linspace(0.3,1, 8)
9
10  my_c1 = ec(f=my_f, d =0.1, c0=100)
11  my_c2 = ec(f=my_f, d=1, c0=100)
12  my_c3 = ec(f=my_f, d=2, c0=100)
13
14  fig, ax = plt.subplots()
15  ax.plot(my_f, my_c1, label="Eq cryst. D = 0.1")
16  ax.plot(my_f, my_c2, label="Eq cryst. D = 1")
17  ax.plot(my_f, my_c3, label="Eq cryst. D = 2")
18
19  ax.set_xlabel('F')
20  ax.set_ylabel('C [ppm]')
21  ax.legend()
```

Listing 3.11 Exploring Eq. (3.1) for various values of D

```
1   import matplotlib.pyplot as plt
2   import numpy as np
3
4   def ec(f, d, c0):
5       c1 = c0/(d * (1-f) + f)
6       return c1
7
8   my_f = np.linspace(0.3,1, 8)
9
10  d = [0.1, 1, 2]
11
```

```
12  fig, ax = plt.subplots()
13
14  for my_d in d:
15      my_c = ec(f=my_f, d=my_d, c0=100)
16      ax.plot(my_f, my_c, label='Eq cryst. D = ' + str(my_d))
17
18
19  ax.set_xlabel('F')
20  ax.set_ylabel('C [ppm]')
21  ax.legend()
```

Listing 3.12 Using a loop to exploring Eq. (3.1) for different values of D

NumPy offers many other ways to define a 1D array. For example, the *np.arange (start, stop, step)* function provides a similar way to obtain a 1D array (see code Listing 3.10).

To investigate Eq. (3.1) for different values of D, we could proceed as done in code Listing 3.11, where we define three models for different values of D (lines 10–12). We then plot the results in a single diagram (lines 15–17) generated at line 14.

Unfortunately, the code Listing 3.11, although easy to understand for a novice, is neither elegant nor efficient. The code Listing 3.12 shows how to obtain the same results by using a loop instead of defining each model separately.

3.3 Quick Intro to Spatial Data Representation

Visualizing spatial data is a fundamental task in geology and has applications in many fields such as geomorphology, hydrology, volcanology, and geochemistry, to cite just a few.

This section outlines a simple task to allow us to become familiar with spatial data. Specifically, we shall import a data elevation model (DEM) stored in a .csv file and display each point using a color proportional to the elevation value. A .csv file is a text file containing data separated by a delimiter such as a comma, tab, or semicolon. To begin, we evaluate the data set stored in the file DEM.csv (see Fig. 3.9), which consists of four columns: a unique index, the elevation, the x coordinate, and the y coordinate. The data set is from the Umbria region in Italy.

Consider now the code Listing 3.13. Lines 1 and 2 import the pandas library and the matplotlib.pyplot subpackage, which are collections of functions and methods to manage and plot scientific data. The command $pd.read\_csv()$ at line 5 imports the .csv file "DEM.csv," creating a new Dataframe (i.e., a table) named *my_data*. *my_data* now contains four columns named *POINTID, ELEVATION, X_LOC*, and *Y_LOC*. The *POINTID* column contains the unique identifiers, the *ELEVATION* column contains the elevation, and the *X_LOC* and *Y_LOC* columns contain the (x, y) coordinates. The command *plt.subplots()* at line 6 generates a figure containing a single axis. Finally, the command *ax.scatter()* at line 7 creates a scatter plot filling

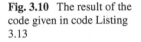

```
1   POINTID,ELEVATION,X_LOC,Y_LOC  ←─────────── Column names
2   1,594.9099731,2306636.664,4832459.757 ◄──────── First row
3   2,1041.869995,2294636.664,4831959.757
4   3,1022.51001,2295136.664,4831959.757 ─────────── Delimiter
5   4,647.7299805,2305136.664,4831959.757
6   5,634.6699829,2306636.664,4831959.757
7   6,649.2800293,2307136.664,4831959.757
8   7,909.7600098,2294136.664,4831459.757
9   8,945.9000244,2294636.664,4831459.757
10  9,979.1699829,2295136.664,4831459.757
11  10,715.2000122,2304636.664,4831459.757
12  11,623.0599976,2305136.664,4831459.757
13  12,640.5,2305636.664,4831459.757
14  13,565.4099731,2306136.664,4831459.757
15  14,660.9500122,2306636.664,4831459.757
16  15,623.1799927,2307136.664,4831459.757
```

Fig. 3.9 Contents of the DEM.csv comma-delimited file (http://www.umbriageo.regione.umbria. it)

Fig. 3.10 The result of the code given in code Listing 3.13

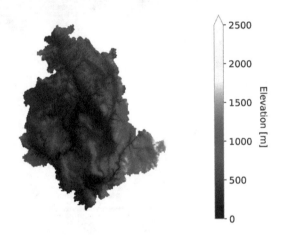

each point defined by X_LOC, Y_LOC with a color proportional to *ELEVATION* (see Fig. 3.10). The argument *cmap='hot'* of *ax.scatter()* at line 7 of code Listing 3.13 sets the colorbar to "hot." In this case, the lowest and the highest values of the colorbar, correspond to black and white, respectively. Intermediate colors mirror the sequence of optical emission from a blackbody becoming progressively hotter (Fig. 3.10). Lines 13–15 provide instructions to plot the colorbar (line 13), set the colorbar label (line 14), and set the color of the colorbar edges (line 15). Figure 3.11 shows the result of code Listing 3.13 with *cmap* equal to "plasma."

Figure 3.12 shows the range of colormaps available in matplotlib.

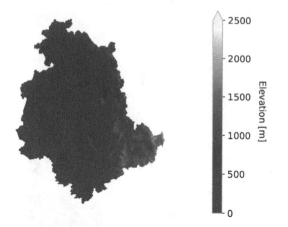

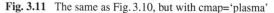

Fig. 3.11 The same as Fig. 3.10, but with cmap='plasma'

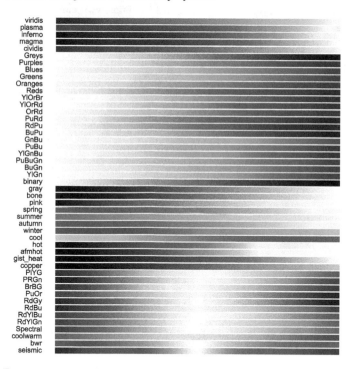

Fig. 3.12 Examples of colormaps available from matplotlib

```
1  import pandas as pd
2  import matplotlib.pyplot as plt
3  from matplotlib import cm
4
5  my_data = pd.read_csv('DEM.csv')
6  fig, ax = plt.subplots()
7  ax.scatter(x = my_data.X_LOC.values,
8             y = my_data.Y_LOC.values,
9             c=my_data.ELEVATION.values,
10            s=2, cmap='hot', linewidth=0, marker='o')
11 ax.axis('equal')
12 ax.axis('off')
13 colorbar = fig.colorbar(cm.ScalarMappable(cmap='hot'), extend='max
      ', ax=ax)
14 colorbar.set_label('Elevation [m]', rotation=270, labelpad=20)
15 colorbar.outline.set_edgecolor('Grey')
```

Listing 3.13 Importing a data elevation model (DEM) stored in a .csv file and displaying the data as a scatter plot

Part II
Describing Geological Data

Chapter 4
Graphical Visualization of a Geological Data Set

4.1 Statistical Description of a Data Set: Key Concepts

Ross [35] states that "statistics is the art of learning from data. It is concerned with the collection of data, their subsequent description, and their analysis, which often leads to the drawing of conclusions." This section provides some basic definitions for a proficient description of a geological data set. Data visualization is of paramount importance to understand data [37], so visualizing data should always come before any advanced statistical modeling [24, 37].

Population: The population is the set of all elements of interest. As an example, suppose we collect the strikes and dips of planar features in a selected area (e.g., bedding planes, foliation planes, fold axial planes, fault planes, and joints). The population of the strikes is the set of all strikes (e.g., for a specific feature). Typically, the whole population cannot be measured, so we must analyze a restricted sample of the population [35].

Sample: A subgroup of the population to be studied in detail is called a sample [35]. Examples are the set of measurements of strikes, spring discharge rates, or the acquisition of CO_2 flow rates for selected locations in volcanic areas. In geology, a piece of rock to be analyzed is also called a sample because it derives from the sampling of a specific rock formation (i.e., the population).

Discrete and continuous data: Discrete data can only take on specific values. An example of discrete data is the number of springs in a specific area. Data are continuous when they can take any value within a range. The results of whole rock analyses and the measurements of flow discharge rates for springs are examples of continuous data [35].

Frequency distribution of a sample: The frequency distribution of a sample is a representation that presents the number of observations within a given interval. It can take either tabular or graphical form [35].

M. Petrelli, *Introduction to Python in Earth Science Data Analysis*,
Springer Textbooks in Earth Sciences, Geography and Environment,
https://doi.org/10.1007/978-3-030-78055-5_4

4.2   Visualizing Univariate Sample Distributions

Histograms

A histogram is a bar-graph containing parallel adjacent bars whose height represents a quantity of interest. It provides a qualitative description of an univariate sample distribution. The vertical axis of a histogram diagram can represent either absolute class frequencies, relative class frequencies, or probability densities. The intervals (i.e., bins) are contiguous and are often of equal size, although this is not required.

The visual inspection of a histogram diagram provides significant information, including (1) the degree of symmetry of the distribution; (2) its spread; (3) the presence of one or more classes characterized by high frequencies; (4) the occurrence of gaps; and (5) the presence of outliers.

The Python statement *matplotlib.axes.Axes.hist()* provides a flexible way to generate and draw histograms. For example, consider code Listing 4.1: Lines 1 and 2 import the pandas library and the matplotlib.pyplot module, respectively. Line 4 defines a DataFrame (i.e., *my_dataset*) by importing the "Supp_traces" spreadsheet from the "Smith_glass_post_NYT_data.xlsx" file. Line 7 generates a new Figure containing a single Axes. Line 8 plots the histogram for column Zr in *my_dataset*.

The arguments *bins* define (1) the number of bins (an integer) and (2) the bin edges (a sequence). In our specific case, *bins = 'auto'* uses a matplotlib internal method to estimate the optimal number of bins (Fig. 4.1).[1]

```
1  import pandas as pd
2  import matplotlib.pyplot as plt
3
4  my_dataset = pd.read_excel(
5      'Smith_glass_post_NYT_data.xlsx', sheet_name='Supp_traces')
6
7  fig, ax = plt.subplots()
8  ax.hist(my_dataset.Zr, bins='auto', edgecolor='black', color='tab:blue
       ', alpha=0.8)
9  ax.set_xlabel('Zr [ppm]')
10 ax.set_ylabel('Counts')
```

Listing 4.1 Plotting a histogram distribution using absolute frequencies in Python

The arguments *color* and *edgecolor* define the color of the bar filling and of the bar edges, respectively. Finally, the argument *alpha* defines the transparency. More details on how to customize an histogram diagram in matplotlib can be found in the official documentation.[2]

---

[1] https://numpy.org/doc/stable/reference/generated/numpy.histogram_bin_edges.html.

[2] https://matplotlib.org/stable/api/_as_gen/matplotlib.axes.Axes.hist.html.

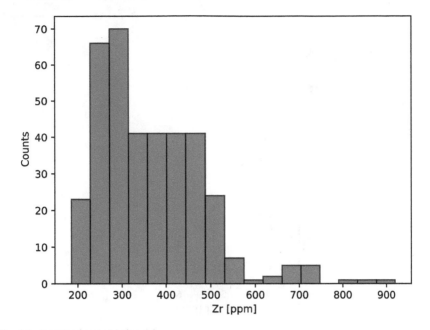

Fig. 4.1 Result of code Listing 4.1

```
1  import pandas as pd
2  import matplotlib.pyplot as plt
3
4  my_dataset = pd.read_excel(
5      'Smith_glass_post_NYT_data.xlsx', sheet_name='Supp_traces')
6
7  fig, ax = plt.subplots()
8  ax.hist(my_dataset.Zr, bins='auto', edgecolor='black', color='tab:blue'
         , alpha=0.8, density=True)
9  ax.set_xlabel('Zr [ppm]')
10 ax.set_ylabel('Probability Density')
```

Listing 4.2 Plotting a histogram distribution as a probability density in Python

Code Listing 4.2 performs the same operations as code Listing 4.1 but adds the instruction *density = True* at line 8. With *density = True*, the *y* axis reports a probability density (Fig. 4.2). In this case, the area under the entire histogram (i.e., the integral) will sum to unity. This is achieved by dividing the absolute frequencies by the bin widths. The use of probability densities correspond to a first attempt to approximate a probability distribution, which is described in Chap. 9.

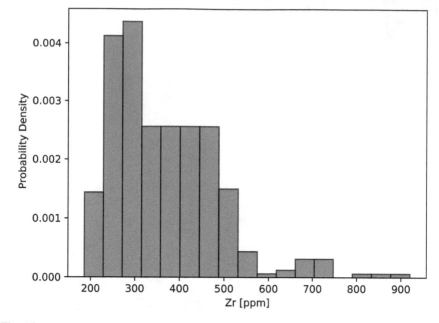

Fig. 4.2 Result of code Listing 4.2

Plot of a cumulative distribution

A cumulative distribution function (also known as a cumulative density function) of a distribution, evaluated at the value x, gives the probability to get values less than or equal to x. Code Listing 4.3 shows how to plot a cumulative distribution using *hist()*. It consists of adding the argument *cumulative = 1* or *cumulative = True* to the *hist()* instruction. The parameter *histtype='step'* prevents the area below the cumulative distribution from being filled. Finally, the parameters *linewidth* and *color* define the line width and color, respectively (Fig. 4.3).

```
1  fig, ax = plt.subplots()
2  ax.hist(my_dataset.Zr, bins='auto', density=True, histtype='step',
       linewidth=2, cumulative=1, color='tab:blue')
3  ax.set_xlabel('Zr [ppm]')
4  ax.set_ylabel('Likelihood of occurrence')
```

Listing 4.3 Plotting a cumulative distribution in Python

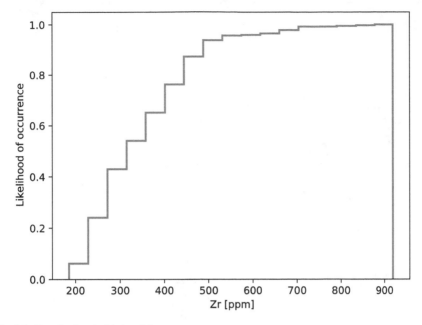

Fig. 4.3 Result of code Listing 4.3

4.3 Preparing Publication-Ready Binary Diagrams

Subplots

Many options are available to create multiple subplots in matplotlib. One of the easiest is to create an empty figure (i.e., *fig = plt.figure()*), then add multiple Axes (i.e., subplots) by using the method $fig.add\_subplot(nrows, ncols, index)$. The parameters (*nrows*), (*ncols*), and *index* give the numbers of rows and columns (*ncols*) and the positional index, respectively. The *index* starts at 1 in the upper-left corner and increases upon moving to the right.

To better understand, consider code Listing 4.4: Line 1 imports the matplotlib. pyplot module, and line 3 generates a new empty figure (i.e., fig). From line 5, we start creating and plotting a grid of diagrams (i.e., three columns and two rows) using $fig.add\_subplot()$ (i.e., lines 5, 9, 13, 17, 21, and 25). In the middle of each diagram, we use the command *text()* to insert text that gives *nrows*, *ncols*, and *index*. Finally, the command $tight\_layout()$ automatically adjusts subplot parameters so that the subplot(s) fits into the figure area (Fig. 4.4). Not using $tight\_layout()$ some elements of the diagram may overlap.

```
1   import matplotlib.pyplot as plt
2
3   fig = plt.figure()
4   # index 1
5   ax1 = fig.add_subplot(2, 3, 1)
6   ax1.text(0.5, 0.5, str((2, 3, 1)), fontsize=18, ha='center')
7
8   # index 2
9   ax1 = fig.add_subplot(2, 3, 2)
10  ax1.text(0.5, 0.5, str((2, 3, 2)), fontsize=18, ha='center')
11
12  # index 3
13  ax1 = fig.add_subplot(2, 3, 3)
14  ax1.text(0.5, 0.5, str((2, 3, 3)), fontsize=18, ha='center')
15
16  # index 4
17  ax1 = fig.add_subplot(2, 3, 4)
18  ax1.text(0.5, 0.5, str((2, 3, 4)), fontsize=18, ha='center')
19
20  # index 5
21  ax1 = fig.add_subplot(2, 3, 5)
22  ax1.text(0.5, 0.5, str((2, 3, 5)), fontsize=18, ha='center')
23
24  # index6
25  ax1 = fig.add_subplot(2, 3, 6)
26  ax1.text(0.5, 0.5, str((2, 3, 6)), fontsize=18, ha='center')
27
28  plt.tight_layout()
```

Listing 4.4 Subplots with matplotlib

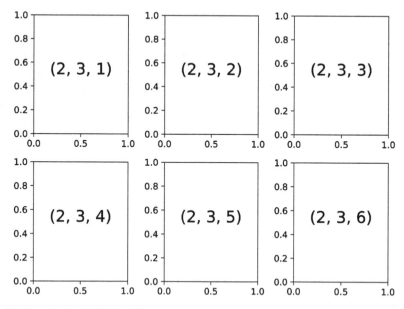

Fig. 4.4 Result of code Listing 4.4

As already noted, code Listing 4.4, although easy to understand for a Python novice, is neither elegant nor efficient. The same results may be obtained by using a for loop, as done in code Listing 4.5.

```
1  import matplotlib.pyplot as plt
2
3  fig = plt.figure()
4
5  for i in range(1, 7):
6      ax = fig.add_subplot(2, 3, i)
7      plt.text(0.5, 0.5, str((2, 3, i)), fontsize=18, ha='center')
8
9  plt.tight_layout()
```

Listing 4.5 Subplots created by using a *for* loop

Markers

The option *markers* in scatter diagrams or other plots determines the shape of the symbol used to identify samples in the diagram. Code Listing 4.6 and Fig. 4.5 show how to use the parameter *marker*. Table 4.1 presents a list of markers available in Python.

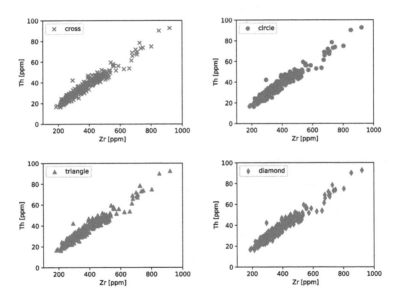

Fig. 4.5 Result of code Listing 4.6. The codes to change markers are reported in Table 4.1

```
1  fig = plt.figure()
2
3  ax1 = fig.add_subplot(2, 2, 1)
4  ax1.scatter(my_dataset.Zr, my_dataset.Th, marker='x', label="cross")
5  ax1.set_xlabel("Zr [ppm]")
6  ax1.set_ylabel("Th [ppm]")
7  ax1.set_xlim([100, 1000])
8  ax1.set_ylim([0, 100])
9  ax1.legend()
10
11  ax2 = fig.add_subplot(2, 2, 2)
12  ax2.scatter(my_dataset.Zr, my_dataset.Th, marker='o', label="circle")
13  ax2.set_xlabel("Zr [ppm]")
14  ax2.set_ylabel("Th [ppm]")
15  ax2.set_xlim([100, 1000])
16  ax2.set_ylim([0, 100])
17  ax2.legend()
18
19  ax3 = fig.add_subplot(2, 2, 3)
20  ax3.scatter(my_dataset.Zr, my_dataset.Th, marker='^', label="triangle")
21  ax3.set_xlabel("Zr [ppm]")
22  ax3.set_ylabel("Th [ppm]")
23  ax3.set_xlim([100, 1000])
24  ax3.set_ylim([0, 100])
25  ax3.legend()
26
27  ax4 = fig.add_subplot(2, 2, 4)
28  ax4.scatter(my_dataset.Zr, my_dataset.Th, marker='d', label="diamond")
29  ax4.set_xlabel("Zr [ppm]")
30  ax4.set_ylabel("Th [ppm]")
31  ax4.set_xlim([100, 1000])
32  ax4.set_ylim([0, 100])
33  ax4.legend()
34
35  fig.tight_layout()
```

Listing 4.6 Setting markers in scatter diagrams

Table 4.1 Marker codes for matplotlib scatter and plot diagrams

Marker	Symbol	Marker	Symbol	Marker	Symbol
.	●	o	●	v	▼
^	▲	<	◀	>	▶
1	Y	2	⅄	4	ⵉ
4	ⵋ	8	●	s	■
p	⬟	h	⬢	H	⬡
+	+	x	×	D	◆
d	◆	|	|	|	|	—	—

Marker dimensions

In scatter diagrams, the marker size can be dictated by the option *s*. Code Listing 4.7 and Fig. 4.6 show how to set marker size.

```
1   fig = plt.figure()
2
3   ax1 = fig.add_subplot(2, 2, 1)
4   plt.scatter(my_dataset.Zr, my_dataset.Th, marker='o', s=10, label="size
        10")
5   ax1.set_xlabel("Zr [ppm]")
6   ax1.set_ylabel("Th [ppm]")
7   ax1.set_xlim([100, 1000])
8   ax1.set_ylim([0, 100])
9   ax1.legend()
10
11  ax2 = fig.add_subplot(2, 2, 2)
12  ax2.scatter(my_dataset.Zr, my_dataset.Th, marker='o', s=50, label="size
        50")
13  ax2.set_xlabel("Zr [ppm]")
14  ax2.set_ylabel("Th [ppm]")
15  ax2.set_xlim([100, 1000])
16  ax2.set_ylim([0, 100])
17  ax2.legend()
18
19  ax3 = fig.add_subplot(2, 2, 3)
20  ax3.scatter(my_dataset.Zr, my_dataset.Th, marker='o', s=100, label="
        size 100")
21  ax3.set_xlabel("Zr [ppm]")
22  ax3.set_ylabel("Th [ppm]")
23  ax3.set_xlim([100, 1000])
24  ax3.set_ylim([0, 100])
25  ax3.legend()
26
27  ax4 = fig.add_subplot(2, 2, 4)
28  ax4.scatter(my_dataset.Zr, my_dataset.Th, marker='o', s=200, label="
        size 200")
29  ax4.set_xlabel("Zr [ppm]")
30  ax4.set_ylabel("Th [ppm]")
31  ax4.set_xlim([100, 1000])
32  ax4.set_ylim([0, 100])
33  ax4.legend()
34
35  fig.tight_layout()
```

Listing 4.7 Setting marker sizes in scatter diagrams

Marker colors

The color of both the marker edge and body of markers may be defined in scatter diagrams by using the options *edgecolor* and *c*, respectively (see code Listing 4.8 and Fig. 4.9). The options *c* and *edgecolor* can be a sequence of colors (e.g., one for each symbol of the diagram) or a single value. In the latter case, the same color is used for all symbols in the diagram. Color values can be specified in different ways. Examples include hexadecimal RGB values (e.g., "#8B0000"), letters or names (see Fig. 4.7, taken from the official documentation of matplotlib[3]), and gray scale levels (i.e., a value from 0 to 1, where 0 is black and 1 is white).

---

[3] https://matplotlib.org/examples/color/named_colors.html.

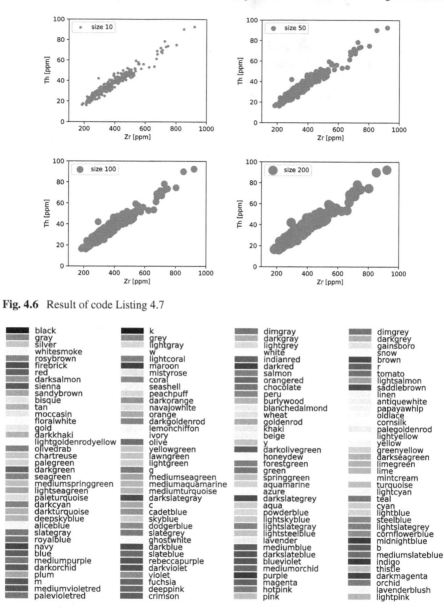

Fig. 4.6 Result of code Listing 4.7

black	k	dimgray	dimgrey
gray	grey	darkgray	darkgrey
silver	lightgray	lightgrey	gainsboro
whitesmoke	w	white	snow
rosybrown	lightcoral	indianred	brown
firebrick	maroon	darkred	r
red	mistyrose	salmon	tomato
darksalmon	coral	orangered	lightsalmon
sienna	seashell	chocolate	saddlebrown
sandybrown	peachpuff	peru	linen
bisque	darkorange	burlywood	antiquewhite
tan	navajowhite	blanchedalmond	papayawhip
moccasin	orange	wheat	oldlace
floralwhite	darkgoldenrod	goldenrod	cornsilk
gold	lemonchiffon	khaki	palegoldenrod
darkkhaki	ivory	beige	lightyellow
lightgoldenrodyellow	olive	y	yellow
olivedrab	yellowgreen	darkolivegreen	greenyellow
chartreuse	lawngreen	honeydew	darkseagreen
palegreen	lightgreen	forestgreen	limegreen
darkgreen	g	green	lime
seagreen	mediumseagreen	springgreen	mintcream
mediumspringgreen	mediumaquamarine	aquamarine	turquoise
lightseagreen	mediumturquoise	azure	lightcyan
paleturquoise	darkslategray	darkslategrey	teal
darkcyan	c	aqua	cyan
darkturquoise	cadetblue	powderblue	lightblue
deepskyblue	skyblue	lightskyblue	steelblue
aliceblue	dodgerblue	lightslategray	lightslategrey
slategray	slategrey	lightsteelblue	cornflowerblue
royalblue	ghostwhite	lavender	midnightblue
navy	darkblue	mediumblue	b
blue	slateblue	darkslateblue	mediumslateblue
mediumpurple	rebeccapurple	blueviolet	indigo
darkorchid	darkviolet	mediumorchid	thistle
plum	violet	purple	darkmagenta
m	fuchsia	magenta	orchid
mediumvioletred	deeppink	hotpink	lavenderblush
palevioletred	crimson	pink	lightpink

Fig. 4.7 Named colors, taken from the official documentation of matplotlib

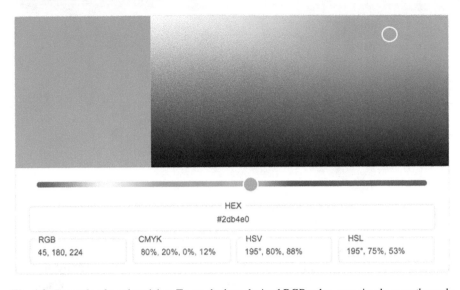

Fig. 4.8 Example of a color picker. To get the hexadecimal RGB value, you simply copy the code in the HEX box

The greatest flexibility with colors is attained by using the hexadecimal RGB values, also known as HEX codes. A HEX code consists of the symbol "#" followed by six digits, which is a sequence of three hexadecimal values ranging from 00 to FF (i.e., from 0 to 255 in decimal notation). The first, second, and third hexadecimal values represent the red, green, and blue components of the color, respectively. At first sight, the HEX notation may appear hard to use; however, you only need to use a "color picker" to select the color of your choice and get the appropriate HEX code. Figure 4.8 shows the color picker provided by Google.

```
1  fig = plt.figure()
2
3  ax1 = fig.add_subplot(2, 2, 1)
4  ax1.scatter(my_dataset.Zr, my_dataset.Th, marker='o', s=60, c='#8B0000'
       , edgecolor='#000000', label="example using hex RGB colors")
5  ax1.set_xlabel("Zr [ppm]")
6  ax1.set_ylabel("Th [ppm]")
7  ax1.set_xlim([100, 1000])
8  ax1.set_ylim([0, 100])
9  ax1.legend()
10
11 ax2 = fig.add_subplot(2, 2, 2)
12 ax2.scatter(my_dataset.Zr, my_dataset.Th, marker='o', s=60,  c='r',
       edgecolor='k', label="example using color letters")
13 ax2.set_xlabel("Zr [ppm]")
14 ax2.set_ylabel("Th [ppm]")
15 ax2.set_xlim([100, 1000])
16 ax2.set_ylim([0, 100])
17 ax2.legend()
18
19 ax3 = fig.add_subplot(2, 2, 3)
20 ax3.scatter(my_dataset.Zr, my_dataset.Th, marker='o', s=60, c='blue',
       edgecolor='black', label="example using color names")
```

```
21   ax3.set_xlabel("Zr [ppm]")
22   ax3.set_ylabel("Th [ppm]")
23   ax3.set_xlim([100, 1000])
24   ax3.set_ylim([0, 100])
25   ax3.legend()
26
27   ax4 = fig.add_subplot(2, 2, 4)
28   ax4.scatter(my_dataset.Zr, my_dataset.Th, marker='o', s=60, c='0.4',
          edgecolor='0', label="example using color gray levels")
29   ax4.set_xlabel("Zr [ppm]")
30   ax4.set_ylabel("Th [ppm]")
31   ax4.set_xlim([100, 1000])
32   ax4.set_ylim([0, 100])
33   ax4.legend()
34
35   fig.tight_layout()
```

Listing 4.8 Setting the color of marker edges and body of markers

Managing legends

The legend is a fundamental element of a diagram, often providing the key notation
needed decipher the information presented in a plot. We have already seen at the
beginning of Chap. 3 how to add a legend to a plot by using the *ax.legend()* command.
Remember that *ax.legend()* automatically creates a legend entry for each labeled
element in the diagram.

We now customize the legend by setting its position and adding a title. The
loc parameter sets the legend position in the diagram. Allowed entries for *loc*
are 'best', 'upper right', 'upper left', 'lower left', 'lower right', 'center left', 'center

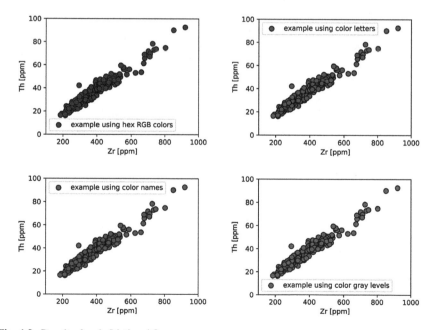

Fig. 4.9 Result of code Listing 4.8

right','lower center','upper center', and'center'. The *loc* parameter can contain the coordinates of the lower-left corner of the legend, as shown in code Listing 4.9, with the result displayed in Fig. 4.10. If not specified, the loc parameter assumes the'best' option, meaning that it searches for the minimum overlap with other graphical elements.

The title parameter adds a title to a legend with title_fontsize defining its font dimension.

Also, *frameon* (True or False), *ncol* (an integer), and *framealpha* (from 0 to 1) define a frame, its transparency, and the numbers of columns, respectively (see code Listing 4.10 and Fig. 4.11).

```python
import pandas as pd
import matplotlib.pyplot as plt

myDataset1 = pd.read_excel('Smith_glass_post_NYT_data.xlsx', sheet_name
    ='Supp_traces')

x = myDataset1.Zr
y = myDataset1.Th


loc_parameters = ['upper right' , 'upper left', 'lower left', 'lower
    right','center'    ,'center left']

fig = plt.figure(figsize=(8,4))
for i in range(len(loc_parameters)):
    ax = fig.add_subplot(2,3,i+1)
    ax.scatter(x, y, marker = 's', color = '#c7ddf4', edgecolor = '
        #000000', label="loc = " + loc_parameters[i])
    ax.set_xlabel("Zr [ppm]")
    ax.set_ylabel("Th [ppm]")
    ax.legend(loc=loc_parameters[i])

fig.tight_layout()
```

Listing 4.9 Customizing legend position using the *loc* parameter

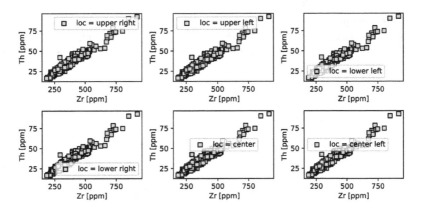

Fig. 4.10 Result of code Listing 4.9

```
1  import pandas as pd
2  import matplotlib.pyplot as plt
3
4  my_dataset = pd.read_excel(
5      'Smith_glass_post_NYT_data.xlsx', sheet_name='Supp_traces')
6
7  my_dataset1 = my_dataset[my_dataset.Epoch == 'one']
8  my_dataset2 = my_dataset[my_dataset.Epoch == 'two']
9
10 fig = plt.figure()
11 ax1 = fig.add_subplot(2, 1, 1)
12 ax1.scatter(my_dataset1.Zr, my_dataset1.Th, marker='s', color='#c7ddf4'
       , edgecolor='#000000', label="First Epoch")
13 ax1.scatter(my_dataset2.Zr, my_dataset2.Th, marker='o', color='#ff464a
       ', edgecolor='#000000', label="Second Epoch")
14 ax1.set_xlabel("Zr [ppm]")
15 ax1.set_ylabel("Th [ppm]")
16 ax1.legend(loc='upper left', framealpha=1, frameon=True, title="Age <
       15 ky", title_fontsize=10)
17
18 ax2 = fig.add_subplot(2, 1, 2)
19 ax2.scatter(my_dataset1.Zr, my_dataset1.Th, marker='s', color='#c7ddf4'
       , edgecolor='#000000', label="First Epoch")
20 ax2.scatter(my_dataset2.Zr, my_dataset2.Th, marker='o', color='#ff464a
       ', edgecolor='#000000', label="Second Epoch")
21 ax2.set_xlabel("Zr [ppm]")
22 ax2.set_ylabel("Th [ppm]")
23 ax2.legend(frameon=False, loc='lower right', ncol=2, title="Age < 15 ky
       ", title_fontsize=10)
24
25 fig.tight_layout()
```

Listing 4.10 Customizing legend parameters

Rounding decimals, text formatting, symbols, and special characters

Reporting data in diagrams (e.g., in the legend or as an annotation), often requires
rounding a number or formatting a string. In such cases, the *.format()* method is
a flexible and useful tool: it allows positional injection of variables (i.e., numbers
or strings) within strings and value formatting. To insert a variable into a string, it
uses a placeholder (i.e., {}). In addition, it allows you to format dates, times, and
numbers and to round decimals. To better understand, consider code Listing 4.11,
which presents practical examples of the use of *.format()*. At lines 5 and 7, *.format()*
is used to insert the two variables *name* and *surname* at specific positions in the text.
Also, lines 12–15 shows how to insert a value (i.e., Archimedes' constant) and round
it to a specific number of digits.

Also, code Listing 4.12 provides additional examples on the use of plus and minus
in *.format()* (lines 5 and 6), reporting numbers as a percent (line 12), and scientific
notation (lines 18 and 19).

```
1  # Introductory examples
2  name = 'Maurizio'
3  surname = 'Petrelli'
4  print('-----------------------------------------------------------')
5  print('My name is {}'.format(name))
6  print('-----------------------------------------------------------')
7  print('My name is {} and my surname is {}'.format(name, surname))
8  print('-----------------------------------------------------------')
```

```
9   # Decimal Number formatting
10  PI = 3.14159265358979323846
11  print('-------------------------------------------------------')
12  print("The 2 digit Archimedes' constant is equal to {:.2f}".format(PI))
13  print("The 3 digit Archimedes' constant is equal to {:.3f}".format(PI))
14  print("The 4 digit Archimedes' constant is equal to {:.4f}".format(PI))
15  print("The 5 digit Archimedes' constant is equal to {:.5f}".format(PI))
16  print('-------------------------------------------------------')
17
18  '''Results
19  -------------------------------------------------------
20  My name is Maurizio
21  -------------------------------------------------------
22  My name is Maurizio and my surname is Petrelli
23  -------------------------------------------------------
24  -------------------------------------------------------
25  The 2 digit Archimedes' constant is equal to 3.14
26  The 3 digit Archimedes' constant is equal to 3.142
27  The 4 digit Archimedes' constant is equal to 3.1416
28  The 5 digit Archimedes' constant is equal to 3.14159
29  -------------------------------------------------------
30  '''
```

Listing 4.11   Familiarize yourself with *.format()*

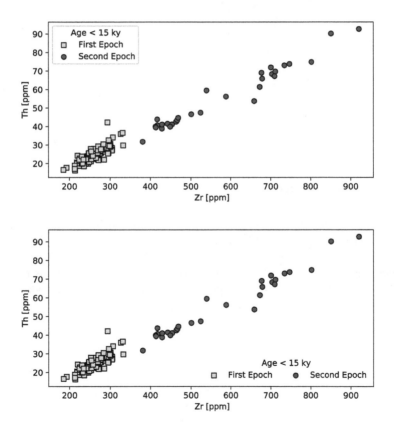

Fig. 4.11   Result of code Listing 4.10

In addition, the escape character "\" is used to insert illegal characters in a string (e.g., " or " when they define the string) or do a specific action (e.g., go to a new line). Table 4.2 and code Listing 4.13 provide some useful examples.

```
1   # Explicit positive and negative reporting
2   a = +5.34352
3   b = -6.3421245
4   print('--------------------------------------------------------')
5   print("The plus symbol is not reported: {:.2f} | {:.2f}".format
        (+5.34352, -6.3421245))
6   print("The plus symbol is reported: {:+.2f} | {:+.2f}".format(a, b))
7   print('--------------------------------------------------------')
8
9   # Reporting as percent
10  c = 0.1558
11  print('--------------------------------------------------------')
12  print("Reporting as percent: {:.1%}".format(c))
13  print('--------------------------------------------------------')
14
15  # Scientific notation
16  d = 6580000000000
17  print('--------------------------------------------------------')
18  print("Scientific notation using e: {:.1e}".format(d))
19  print("Scientific notation using E: {:.1E}".format(d))
20  print('--------------------------------------------------------')
21
22  '''Results
23  --------------------------------------------------------
24  The plus symbol is not reported: 5.34 | -6.34
25  The plus symbol is reported: +5.34 | -6.34
26  --------------------------------------------------------
27  --------------------------------------------------------
28  Reporting as percent: 15.6%
29  --------------------------------------------------------
30  --------------------------------------------------------
31  Scientific notation using e: 6.6e+12
32  Scientific notation using E: 6.6E+12
33  --------------------------------------------------------
34  '''
```

Listing 4.12 More examples of reporting numbers using *.format()*

Our next challenge is to insert symbols and equations into diagrams. In my opinion, the simplest and most direct way to apply text formatting (e.g., superscripts and subscripts), insert symbols (e.g., μ or η), or introduce special characters (e.g., $\pm$) in matplotlib is to use TEX markup. In brief, TEX provides the foundations for LATEX, a high-quality typesetting system. In practice, LATEX is the *de facto* standard for the communication and publication of scientific documents.[4]

---

[4] https://www.latex-project.org.

Table 4.2 Using the escape character \ to insert special characters in a string or to do specific actions

Command	Result	Command	Result	Command	Result
\ n	New Line	\ '	Single Quote	\ "	Double Quote
\textbackslash	\	\ ooo	Octal value	\ xhh	Hex value

```
1   # Go to new line using \n
2   print('-----------------------------------------------------------')
3   print("My name is\nMaurizio Petrelli")
4
5   # Inserting characters using octal values
6   print('-----------------------------------------------------------')
7   print("\100 \136 \137 \077 \176")
8
9   # Inserting characters using hex values
10  print('-----------------------------------------------------------')
11  print("\x23 \x24 \x25 \x26 \x2A")
12  print('-----------------------------------------------------------')
13
14  '''Output:
15  -----------------------------------------------------------
16  My name is
17  Maurizio Petrelli
18  -----------------------------------------------------------
19  @ ^ _ ? ~
20  -----------------------------------------------------------
21  # $ % & *
22  -----------------------------------------------------------
23  '''
```

Listing 4.13 Examples of how to use the escape character \

Since teaching TEX and LATEX is far beyond the scope of the present book, the reader is invited to refer to specialized books on the subject [27, 28, 30]. That said, knowing a few specific rules and notations will greatly improve the quality of our diagrams.

Note that any text element in matplotlib can use advanced formatting, mathematical elements, and symbols. To use TEX in matplotlib, we precede the quotes defining a string with r (i.e., r'this is my string'), and enclose math between dollar signs (i.e., $a+b=c\$$) . Code Listing 4.14 shows an example of how to use the TEX notation to improve the quality of our diagrams (Fig. 4.12).

Table 4.3 provides some common TEX instructions, such as how to apply superscripts and subscripts, insert Greek letters such as μ, η, and π and special characters such as $\pm$ and ∞, or mathematical expressions such as $\int_a^b$.

Table 4.3 Introducing TEX notation in matplotlib. Example: r'$x ^ {2}$' $\rightarrow$ x^2

TEX	Result	TEX	Result	TEX	Result
x^{2}	x^2	x_{2}	x_2	\ pm	$\pm$
\ alpha	α	\ beta	β	\ gamma	γ
\ rho	ρ	\ sigma	σ	\ delta	δ
\ pi	π	\ eta	η	\ mu	μ
\ int	$\int$	\ sum	$\sum$	\ prod	$\prod$
\ leftarrow	$\leftarrow$	\ rightarrow	$\rightarrow$	\ uparrow	$\uparrow$
\ Leftarrow	$\Leftarrow$	\ Rightarrow	$\Rightarrow$	\ Uparrow	$\Uparrow$
\ infty	∞	\ nabla	∇	\ partial	∂
\ neq	$\neq$	\ simeq	$\simeq$	\ approx	$\approx$

```
1   import pandas as pd
2   import matplotlib.pyplot as plt
3   import numpy as np
4
5
6   def my_line(x, m, q):
7       y = m * x + q
8       return y
9
10
11  my_dataset = pd.read_excel('Smith_glass_post_NYT_data.xlsx', sheet_name
        ='Supp_majors', engine='openpyxl')
12
13  my_dataset1 = my_dataset[my_dataset.Epoch == 'one']
14  my_dataset2 = my_dataset[my_dataset.Epoch == 'two']
15
16  x = np.linspace(52.5, 62, 100)
17  y = my_line(x, m=0.3, q=-10.3)
18
19  fig, ax = plt.subplots()
20
21  ax.scatter(my_dataset1.SiO2, my_dataset1.K2O, marker='s', color='#
        c7ddf4', edgecolor='#000000', label=r'$1^{st}$ Epoch')
22  ax.scatter(my_dataset2.SiO2, my_dataset2.K2O, marker='s', color='#
        ff464a', edgecolor='#000000', label=r'$2^{nd}$ Epoch')
23  ax.plot(x, y, color='#342a77')
24
25  ax.annotate(r'What is the 1$\sigma$ for this point?', xy=(47.6, 6.6),
        xytext=(47, 8.8), arrowprops=dict(arrowstyle="->", connectionstyle
        ="arc3"))
26  ax.text(52.4, 5.6, r'$ Na_2O = 0.3 \cdot SiO_2 -10.3$', dict(size=10,
        rotation=33))
27
28  ax.text(53.5, 5.1, r'$ \mu_{SiO_2} = \frac {a_{1}+a_{2}+\cdots +a_{n}}{
        n}$ = ' + '{:.1f} [wt.%]'.format(57.721), dict(size=11.5))
29
30  ax.set_xlabel(r'SiO$_2$ [wt%]')
31  ax.set_ylabel(r'K$_2$O [wt%]')
32
33  ax.legend()
```

Listing 4.14 Using TEX notation in matplotlib

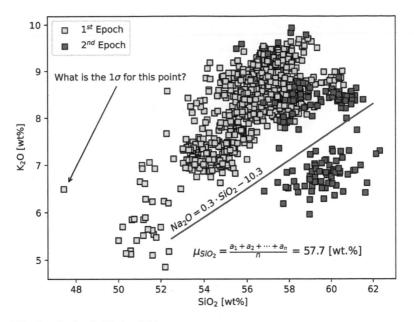

Fig. 4.12 Result of code Listing 4.14

A drawback of this scheme is that adding r to strings precludes the use of *.format()* and \ as escape character. To overcome this problem, you can split the string into sub-strings and then concatenate the sub-strings by using the + symbol, as done in line 28 of code Listing 4.14.

Binary diagrams: *plot()* versus *scatter()*

In the previous sections, we introduced two different methods to visualize geological data in binary diagrams: the *plot()* and *scatter()* functions implemented in the matplotlib sub-package named pyplot. The two methods share many functionalities and can often be used interchangeably. As an example, consider code Listing 4.15, which shows how to plot a binary diagram with square markers (Fig. 4.13).

Of course, *plot()* and *scatter()* differ in some ways (see code Listing 4.16); for example, *plot()* only connects with a line the points defined by a sequence of (x, y) coordinates (see Fig. 4.14). Table 4.4 shows the main parameters available to personalize a *plt.plot()* diagram. However, the *plot()* function is less flexible than *scatter()* for marker sizing and coloring. For each *plot()* declaration, all symbols must be of the same size and color. Conversely, *scatter()* allows you to use different colors and sizes for each marker. For example, Fig. 4.14 shows symbols whose size is proportional to the F parameter and whose color is defined by the color sequence (see code Listing 4.16).

Occasionally, you may need to combine *plot()* and *scatter()*. For example, if you want to plot and connect a sequence of samples with different colors and dimensions,

Table 4.4 Parameters for personalizing a *plot()* diagram

Parameter	Value	Description
alpha	[0,1]	Set transparency
color, c	Color value (e.g., Figs. 4.7 and 4.8)	Set color of the line
fillstyle	{'full', 'left', 'right', 'bottom', 'top', 'none'}	Set marker fill style
linestyle, ls	{'-', '--', '-.', ':', '', (offset, on-off-seq), ...}	Set style of the line
linewidth, lw	Floating point number	Set line width in points
marker	Marker style (e.g., Table 4.1)	Set marker style
markeredgecolor, mec	Color value	Set marker edge color
markeredgewidth, mew	Floating point number	Set marker edge width
markerfacecolor, mfc	Color value	Set marker face color
markersize, ms	Floating point number	Set marker size in points

you could use *scatter()* for symbols and *plot()* for the connecting line. The *zorder* parameter is an integer number defining the stratigraphy of different layers in the diagram. In code Listing 4.16 (lines 33 and 34), *zorder* places symbols above the line (Fig. 4.14).

```
1  import pandas as pd
2  import matplotlib.pyplot as plt
3
4  my_dataset1 = pd.read_excel('Smith_glass_post_NYT_data.xlsx',
       sheet_name='Supp_traces')
5
6  x = my_dataset1.Zr
7  y = my_dataset1.Th
8
9  fig = plt.figure()
10 ax1 = fig.add_subplot(1, 2, 1)
11 ax1.scatter(x, y, marker='s', color='#ff464a', edgecolor='#000000')
12 ax1.set_title("using scatter()")
13 ax1.set_xlabel("Zr [ppm]")
14 ax1.set_ylabel("Th [ppm]")
15 ax2 = fig.add_subplot(1, 2, 2)
16 ax2.plot(x, y, marker='s', linestyle='', color='#ff464a',
       markeredgecolor='#000000')
17 ax2.set_title("using plot()")
18 ax2.set_xlabel("Zr [ppm]")
19 ax2.set_ylabel("Th [ppm]")
20 fig.tight_layout()
```

Listing 4.15 Often, *plot()* and *scatter()* can be used to solve the same tasks.

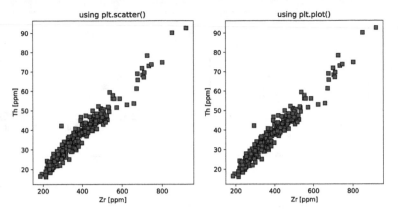

Fig. 4.13 Result of code Listing 4.15

```
1   import matplotlib.pyplot as plt
2   import numpy as np
3
4
5   def ec(f, d, c0):
6       c1 = c0 / (d * (1-f) + f)
7       return c1
8
9
10  my_f = np.linspace(0.1, 1, 10)
11
12  my_c1 = ec(f=my_f, d=0.1, c0=100)
13
14  colors = ['#ff9494', '#cbeaa2', '#d1a396', '#828fc3', '#95b2e5', '#
        e9b8f4', '#f4b8e5', '#b8f4f2', '#c5f4b8', '#f9ca78']
15
16  fig = plt.figure()
17  ax1 = fig.add_subplot(2, 2, 1)
18  ax1.plot(my_f, my_c1, marker='o', linestyle='-', markersize=5)
19  ax1.set_xlabel('F')
20  ax1.set_ylabel('C [ppm]')
21
22  ax2 = fig.add_subplot(2, 2, 2)
23  ax2.scatter(my_f, my_c1, marker='o', s=my_f*150)
24  ax2.set_xlabel('F')
25  ax2.set_ylabel('C [ppm]')
26
27  ax3 = fig.add_subplot(2, 2, 3)
28  ax3.scatter(my_f, my_c1, marker='o', c=colors, s=my_f*150)
29  ax3.set_xlabel('F')
30  ax3.set_ylabel('C [ppm]')
31
32  ax4 = fig.add_subplot(2, 2, 4)
33  ax4.plot(my_f, my_c1, marker='', linestyle='-', zorder=0)
34  ax4.scatter(my_f, my_c1, marker='o', c=colors, s=my_f*150, zorder=1)
35  ax4.set_xlabel('F')
36  ax4.set_ylabel('C [ppm]')
37
38  fig.tight_layout()
```

Listing 4.16 Main differences between *plot()* and *scatter()*

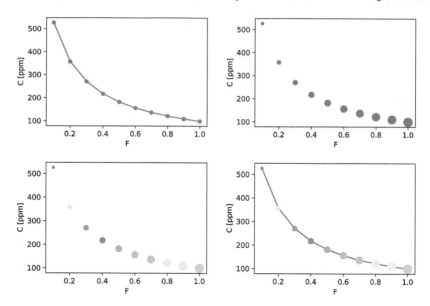

Fig. 4.14   Result of code Listing 4.16

```
1  import pandas as pd
2  import matplotlib.pyplot as plt
3
4  my_dataset = pd.read_excel(
5      'Smith_glass_post_NYT_data.xlsx', sheet_name='Supp_traces')
6
7  epochs = ['one', 'two', 'three', 'three-b']
8  colors = ['#c8b4ba', '#f3ddb3', '#c1cd97', '#e18d96']
9  markers = ['o', 's', 'd', 'v']
10
11 fig, ax = plt.subplots()
12 for (epoch, color, marker) in zip(epochs, colors, markers):
13     my_data = my_dataset[(my_dataset.Epoch == epoch)]
14     ax.scatter(my_data.Zr, my_data.Th, marker=marker, s=50, c=color,
           edgecolor='0', label="Epoch " + epoch)
15
16 ax.set_xlabel("Zr [ppm]")
17 ax.set_ylabel("Th [ppm]")
18 ax.legend(title="Phlegraean Fields \n Age < 15 ky")
```

Listing 4.17   Creating a publication-ready diagram in Python

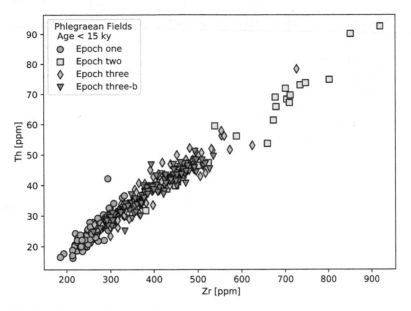

Fig. 4.15 Result of code Listing 4.17

Example of publication-ready diagram

As a final task, we prepare a publication-ready diagram (see code Listing 4.17). Lines 1 and 2 import the pandas library and the matplotlib.pyplot module, respectively. Line 4 imports the Excel file in a DataFrame named *my_dataset*. Lines 7–9 define three sequences named epochs, colors, and markers, respectively. At line 11, we generate a new empty figure, and then line 12 implements a loop that iterates over the sequences of epochs, colors, and markers using the *zip()* function. This enables us to iterate over two or more lists at the same time. Next, line 13 defines a new DataFrame named *my_data* by filtering *my_dataset* using the labels in the Epoch column. At line 14, we add a Zr versus Th scatter diagram of the resulting *my_data* to the figure generated at line 11. Finally, we add axis labels (lines 16 and 17) and a legend with a title (line 18). Note that \ n simply defines a new line in the legend title (Fig. 4.15).

```
1  import pandas as pd
2  import seaborn as sns
3
4  my_dataset = pd.read_excel('Smith_glass_post_NYT_data.xlsx', sheet_name
       ='Supp_traces')
5
6  my_dataset1 = my_dataset[['Ba', 'Zr', 'Th']]
7  sns.pairplot(my_dataset1)
```

Listing 4.18 A first attempt at visualizing multivariate data using *sns.pairplot()*

4.4   Visualization of Multivariate Data: A First Attempt

The function *seaborn.pairplot()* plots pairwise relationships from a data set. By default, this function creates a grid of diagrams where each variable in the data set is shared in the y axis across a single row and in the x axis across a single column. In diagonal diagrams, the *pairplot()* function draws a plot showing the univariate distribution for the variable in that column.[5] Code Listing 4.18 shows how to generate a *pairplot()* diagram using Ba, Zr, and Th. Lines 1 and 2 import the pandas and seaborn libraries, respectively, and line 4 imports an Excel file in a DataFrame named *my_dataset*. Line 6 generates a new DataFrame (i.e., my_dadaset1) by filtering *my_dataset* for the columns Ba, Zr, and Th. More details about the filtering and slicing of a DataFrame are reported in Appendix D. Finally, line 7 generates a pairpolt diagram, which is displayed in Fig. 4.16 by using code Listing 4.18.

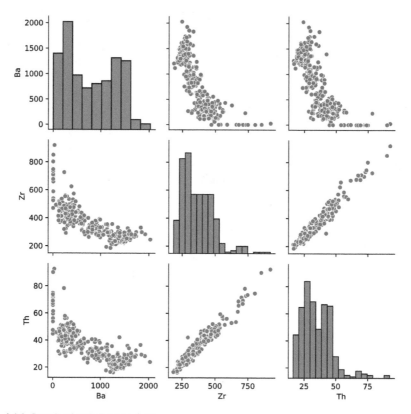

Fig. 4.16 Result of code Listing 4.18

---

[5] https://seaborn.pydata.org/generated/seaborn.pairplot.html.

Chapter 5
Descriptive Statistics 1: Univariate Analysis

5.1 Basics of Descriptive Statistics

Descriptive statistics deals with metrics, tools, and strategies that can be used to summarize a data set. These metrics are extracted from the data and provide information about (1) the location of a data set, sometime defined as the central tendency; (2) the amount of data variation (i.e., the dispersion), and (3) the degree of symmetry (i.e., the skewness). Metrics of the location of a data set are the arithmetic, geometric, and harmonic means. The median and the mode of mono-modal distributions are also measures of the location of a data set. The total spread of a data set is a rough estimate of dispersion. More accurate estimates of the dispersion of a data set are the variance, standard deviation, and inter-quartile range. The skewness of a data set can be measured by quantities such as Pearson's first coefficient of skewness or the Fischer-Pearson coefficient of skewness.

5.2 Location

In descriptive statistics, it is useful to represent an entire data set with a single value describing its location or position. This single value is defined as the central tendency. Mean, median, and mode all fall into this category.

© The Author(s), under exclusive license to Springer Nature Switzerland AG 2021   67
M. Petrelli, *Introduction to Python in Earth Science Data Analysis*,
Springer Textbooks in Earth Sciences, Geography and Environment,
https://doi.org/10.1007/978-3-030-78055-5_5

Mean

The arithmetic mean μ_A is the average of all numbers in a data set and is defined as

$$\mu_A = \bar{z} = \frac{1}{n}\sum_{i=1}^{n} z_i = \frac{z_1 + z_2 + \cdots + z_n}{n}. \tag{5.1}$$

The geometric mean μ_G is a type of mean that indicates the location of a data set by using the product of their values:

$$\mu_G = (z_1 z_2 \cdots z_n)^{\frac{1}{n}}. \tag{5.2}$$

Finally, the harmonic mean μ_H is

$$\mu_H = \frac{n}{\frac{1}{z_1} + \frac{1}{z_2} + \cdots + \frac{1}{z_n}}. \tag{5.3}$$

In the following, when not explicitly specified, the symbol μ is to be understood as the arithmetic mean. Code Listing 5.1 and Fig. 5.1 show one way to get the different means for a specific feature (in our case, the concentration of a chemical element like zirconium, Zr) in the imported data set.

```python
import pandas as pd
from scipy.stats.mstats import gmean, hmean
import matplotlib.pyplot as plt

my_dataset = pd.read_excel('Smith_glass_post_NYT_data.
    xlsx', sheet_name='Supp_traces')

a_mean = my_dataset.Zr.mean()
g_mean = gmean(my_dataset['Zr'])
h_mean = hmean(my_dataset['Zr'])

print ('--------')
print ('arithmetic mean')
print ("{0:.1f} [ppm]".format(a_mean))
print ('--------')

print ('geometric mean')
print ("{0:.1f} [ppm]".format(g_mean))
print ('--------')

print ('harmonic mean')
print ("{0:.1f} [ppm]".format(h_mean))
print ('--------')

fig, ax = plt.subplots()
ax.hist(my_dataset.Zr, bins='auto', density=True,
    edgecolor='k', label='Measurements Hist', alpha=0.8)
```

```
26  ax.axvline(a_mean, color='purple', label='Arithmetic mean
        ', linewidth=2)
27  ax.axvline(g_mean, color='orange', label='Geometric mean
        ', linewidth=2)
28  ax.axvline(h_mean, color='green', label='Harmonic mean',
        linewidth=2)
29  ax.set_xlabel('Zr [ppm]')
30  ax.set_ylabel('Probability density')
31  ax.legend()
32
33  '''
34  Output:
35  -------
36  arithmetic mean
37  365.4 [ppm]
38  -------
39  geometric mean
40  348.6 [ppm]
41  -------
42  harmonic mean
43  333.8 [ppm]
44  -------
45  '''
```

Listing 5.1 Measuring and plotting the mean values of a data set

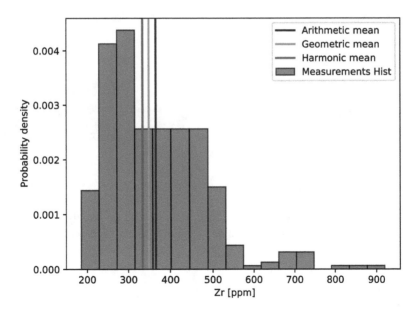

Fig. 5.1 Result of code Listing 5.1

Median

The median Me is the number in the middle of a data set after sorting from the lowest to the highest value (see code Listing 5.2 and Fig. 5.2). Consequently, to obtain the median of a data set, the data must be ordered from smallest to largest. If the number of data values is odd, then the sample median is the middle value in the ordered list; if it is even, then the sample median is the average of the two middle values [35].

```python
1   import pandas as pd
2   import matplotlib.pyplot as plt
3
4   my_dataset = pd.read_excel('Smith_glass_post_NYT_data.
        xlsx', sheet_name='Supp_traces')
5
6   median = my_dataset.Zr.median()
7
8   print ('--------')
9   print ('median')
10  print ("{0:.1f} [ppm]".format(median))
11  print ('--------')
12
13  fig, ax = plt.subplots()
14  ax.hist(my_dataset.Zr, bins=20, density=True, edgecolor='
        k', label="Measurements Hist", alpha=0.8)
15  ax.axvline(median, color='orange', label='Median',
        linewidth=2)
16  ax.set_xlabel('Zr [ppm]')
17  ax.set_ylabel('Probability density')
18  ax.legend()
19
20  '''
21  Output:
22  -------
23  median
24  339.4 [ppm]
25  -------
26  '''
```

Listing 5.2 Measuring and plotting the median of a data set

Mode

The mode Mo of a data set is the value that appears most frequently in the data set [35]. In Python, the mode may be obtained as shown in code Listing 5.3 (Fig. 5.3).

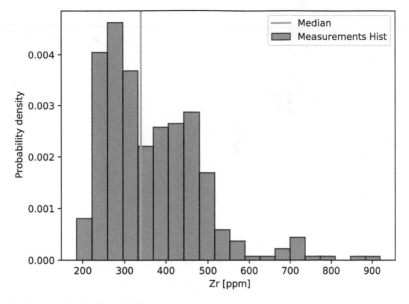

Fig. 5.2 Result of code Listing 5.2

```
1  import pandas as pd
2  import numpy as np
3  import matplotlib.pyplot as plt
4
5  my_dataset = pd.read_excel('Smith_glass_post_NYT_data.
       xlsx', sheet_name='Supp_traces')
6
7  hist, bin_edges = np.histogram(my_dataset['Zr'], bins=
       20, density=True)
8  modal_value = (bin_edges[hist.argmax()] + bin_edges[hist
       .argmax()+1])/2
9
10 print ('modal value: {0:.0f} [ppm]'.format(modal_value))
11
12 fig, ax = plt.subplots()
13 ax.hist(my_dataset.Zr, bins=20, density=True, edgecolor='
       k', label="Measurements Hist", alpha=0.8)
14 ax.axvline(modal_value, color="orange", label="Modal
       value", linewidth=2)
15 ax.set_xlabel('Zr [ppm]')
16 ax.set_ylabel('Probability density')
17 ax.legend()
18
19 '''
20 Output: modal value: 277 [ppm]
21 '''
```

Listing 5.3 Measuring and plotting the mode of a data set

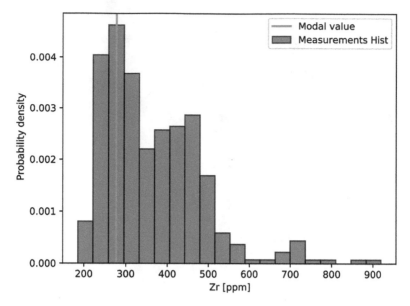

Fig. 5.3 Result of code Listing 5.3

5.3   Dispersion or Scale

After introducing several estimators of the central tendency of a data set, we now consider measures of its variability. The range, variance, and standard deviation are all estimators of the dispersion (i.e., variability) of a data set.

Range

A first gross estimator of the variability of a data set is its range R, which is the difference between the highest and lowest values in a data set (see code Listing 5.4 and Fig. 5.4):

$$R = z_{max} - z_{min}. \tag{5.4}$$

```
1  import pandas as pd
2  import matplotlib.pyplot as plt
3
4  my_dataset = pd.read_excel('Smith_glass_post_NYT_data.
       xlsx', sheet_name='Supp_traces')
5
6  my_range = my_dataset['Zr'].max()- my_dataset['Zr'].min()
7
8  print ('--------')
9  print ('Range')
10 print("{0:.0f}".format(my_range))
11 print ('--------')
12
```

```
13  fig, ax = plt.subplots()
14  ax.hist(my_dataset.Zr, bins=20, density=True, edgecolor='
        k', label='Measurements Hist')
15  ax.axvline(my_dataset['Zr'].max(), color='purple', label=
        'Max value', linewidth=2)
16  ax.axvline(my_dataset['Zr'].min(), color='green', label='
        Min value', linewidth=2)
17  ax.axvspan(my_dataset['Zr'].min(), my_dataset['Zr'].max()
        , alpha=0.1, color='orange', label='Range = ' + "{0:.0
        f}".format(my_range) + ' ppm')
18  ax.set_xlabel('Zr [ppm]')
19  ax.set_ylabel('Probability density')
20  ax.legend()
```

Listing 5.4 Measuring and plotting the range of a data set

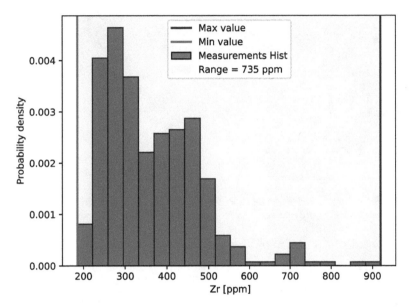

Fig. 5.4 Result of code Listing 5.4

Variance and standard deviation

The variances for population (σ_p^2) and sample (σ_s^2) distributions are defined as

$$\sigma_\mathrm{p}^2 = \frac{\sum_{i=1}^{n}(z_i - \mu)^2}{n}, \tag{5.5}$$

$$\sigma_\mathrm{s}^2 = \frac{\sum_{i=1}^{n}(z_i - \mu)^2}{n-1}. \tag{5.6}$$

The standard deviation σ is the square root of the variance:

$$\sigma_p = \sqrt{\sigma_p^2} = \sqrt{\frac{\sum_{i=1}^{n}(z_i - \mu)^2}{n}}, \tag{5.7}$$

$$\sigma_s = \sqrt{\sigma_s^2} = \sqrt{\frac{\sum_{i=1}^{n}(z_i - \mu)^2}{n} - 1}. \tag{5.8}$$

The variance σ_s^2 and the standard deviation σ_s of a sample distribution can be estimated in pandas as shown in code Listing 5.5.

```
1  import pandas as pd
2  import matplotlib.pyplot as plt
3
4  my_dataset = pd.read_excel('Smith_glass_post_NYT_data.
       xlsx', sheet_name='Supp_traces')
5
6  variance =  my_dataset['Zr'].var()
7  stddev =  my_dataset['Zr'].std()
8
9  print ('-------')
10 print ('Variance')
11 print("{0:.0f} [square ppm]".format(variance))
12 print ('-------')
13 print ('Standard Deviation')
14 print("{0:.0f} [ppm]".format(stddev))
15 print ('-------')
16
17 fig, ax = plt.subplots()
18 ax.hist(my_dataset.Zr, bins= 20, density = True,
       edgecolor='k', label='Measurements Hist')
19 ax.axvline(my_dataset['Zr'].mean()-stddev, color='purple'
       , label=r'mean - 1$\sigma$', linewidth=2)
20 ax.axvline(my_dataset['Zr'].mean()+stddev, color='green',
       label=r'mean + 1$\sigma$', linewidth=2)
21 ax.axvspan(my_dataset['Zr'].mean()-stddev, my_dataset['Zr
       '].mean()+stddev, alpha=0.1, color='orange', label=r'
       mean $\pm$ 1$\sigma$')
22 ax.set_xlabel('Zr [ppm]')
23 ax.set_ylabel('Probability density')
24 ax.legend()
25
26 '''
27 Output:
28 -------
29 Variance
30 14021 [square ppm]
31 -------
32 Standard Deviation
33 118 [ppm]
34 -------
35 '''
```

Listing 5.5 Measuring and plotting the variance and standard deviation of a data set

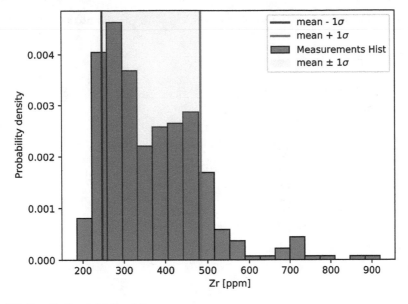

Fig. 5.5 Result of code Listing 5.5

In pandas, to estimate the variance and the standard deviation for an entire population, you need to change the Delta Degrees of Freedom (ddof). By default, the pandas commands *.var()* and *.std()* use *ddof = 1*, normalizing the measurements by $n - 1$. Setting *.var(ddof = 0)* and *.std(ddof = 0)* makes pandas calculate σ_p^2 and σ_p, respectively. Variances and standard deviations can also serve as estimates of NumPy arrays by using the same *.var()* and *.std()* commands. Unlike pandas, the NumPy implementations of *.var()* and *.std()* use *ddof = 0* by default, calculating the population variance σ_p^2 and standard deviation σ_p, respectively.

Inter-quartile range

In descriptive statistics, the inter-quartile range is the difference between the 75th and 25th percentiles, or between the upper and lower quartiles (see code Listing 5.6 and Fig. 5.6). For the meaning of the parameter *interpolation*, please refer to the official documentation.[1]

```
1  import pandas as pd
2  import numpy as np
3  import matplotlib.pyplot as plt
4
5  my_dataset = pd.read_excel('Smith_glass_post_NYT_data.
       xlsx', sheet_name='Supp_traces')
6
7  q1 = np.percentile(my_dataset.Zr, 25, interpolation = '
       midpoint')
```

---

[1] https://numpy.org/doc/stable/reference/generated/numpy.percentile.html.

```
 8  q3 = np.percentile(my_dataset.Zr, 75, interpolation = '
        midpoint')
 9
10  iqr = q3 - q1 # Interquaritle range (IQR)
11
12  print ('--------')
13  print ('Interquaritle range (IQR): {0:.0f} [ppm]'.format(
        iqr))
14  print ('--------')
15
16  fig, ax = plt.subplots()
17  ax.hist(my_dataset.Zr, bins='auto', density=True,
        edgecolor='k', label='Measurements Hist')
18  ax.axvline(q1, color='purple', label='Q1', linewidth=2)
19  ax.axvline(q3, color='green', label='Q3', linewidth=2)
20  ax.axvspan(q1, q3, alpha=0.1, color='orange', label='
        Interquaritle range (IQR)')
21  ax.set_xlabel('Zr [ppm]')
22  ax.set_ylabel('Probability density')
23  ax.legend()
24
25  '''
26  Output:
27  -------
28  Interquaritle range (IQR): 164 [ppm]
29  -------
30  '''
```

Listing 5.6 Measuring and plotting the inter-quartile range of a data set

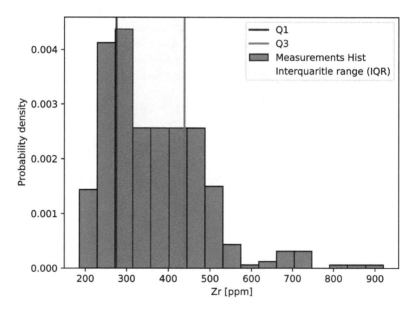

Fig. 5.6 Result of code Listing 5.6

5.4 Skewness

Having introduced various parameters providing information about the central tendency and the variability of a data set, we can now start analyzing the skewness, which reflects the shape of a distribution.

The skewness is a statistical parameter that provides information about the symmetry in a distribution of values. In the case of a symmetric distribution, the mean, median, and mode are the same: $Mo = Me = \mu_A$. Note, however, that the coincidence of these three values, although being a necessary condition for a symmetric distributions, does not guarantee the symmetry of a distribution. Conversely, the non-coincidence of these three parameters indicates a skewed distribution. In particular, when $Mo < Me < \mu_A$ and $\mu_A < Me < Mo$, the distribution is characterized by tails on the right and left side, respectively.

In the specific case of the concentration distribution of Zr, where $Mo < Me < \mu_A$, a tail appears on the right side, as expected (see code Listing 5.7 and Fig. 5.7).

```
1  import pandas as pd
2  import numpy as np
3  import matplotlib.pyplot as plt
4
5  my_dataset = pd.read_excel('Smith_glass_post_NYT_data.
       xlsx', sheet_name='Supp_traces')
6
7  a_mean = my_dataset.Zr.mean()
8
9  median = my_dataset.Zr.median()
10
11 hist, bin_edges = np.histogram(my_dataset['Zr'], bins=20,
       density=True)
12 modal_value = (bin_edges[hist.argmax()] + bin_edges[hist
       .argmax()+1])/2
13
14 fig, ax = plt.subplots()
15 ax.hist(my_dataset.Zr, bins=20, density=True, edgecolor='
       k', label="Measurements Hist")
16 ax.axvline(modal_value, color='orange', label='Modal
       Value', linewidth=2)
17 ax.axvline(median, color='purple', label='Median Value',
       linewidth=2)
18 ax.axvline(a_mean, color='green', label='Arithmetic mean
       ', linewidth=2)
19 ax.set_xlabel('Zr [ppm]')
20 ax.set_ylabel('Probability density')
21 ax.legend()
```

Listing 5.7 A qualitative test of the skewness of a data set

Another parameter providing information about the skewness of a sample distribution is Pearson's first coefficient of skewness, which is given by

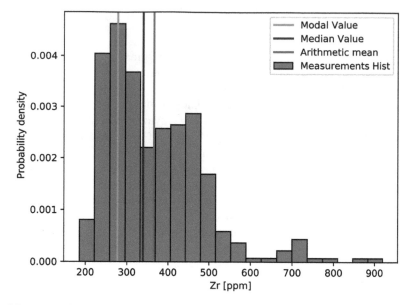

Fig. 5.7 Result of code Listing 5.7

$$\alpha_1 = \frac{(\mu - Mo)}{\sigma_s}. \tag{5.9}$$

A second parameter is the Pearson's second moment of skewness,

$$\alpha_2 = \frac{3(\mu - Me)}{\sigma_s}. \tag{5.10}$$

An additional parameter providing information about the sample skewness is the Fisher-Pearson coefficient of skewness,

$$g_1 = \frac{m_3}{m_2^{3/2}}, \tag{5.11}$$

where

$$m_i = \frac{1}{N} \sum_{n=1}^{N} (x[n] - \mu)^i. \tag{5.12}$$

In Python, the parameters α_1, α_2, and g_1 can be determined as shown in code Listing 5.8.

```
1  import numpy as np
2  from scipy.stats import skew
3
4  a_mean = my_dataset.Zr.mean()
5  median = my_dataset.Zr.median()
```

```
6  hist, bin_edges = np.histogram(my_dataset['Zr'], bins=20,
       density=True)
7  modal_value = (bin_edges[hist.argmax()] + bin_edges[hist.
       argmax()+1])/2
8  standard_deviation = my_dataset['Zr'].std()
9
10 a1 = (a_mean - modal_value) / standard_deviation
11 a2 = 3 * (a_mean - median) / standard_deviation
12 g1 = skew(my_dataset['Zr'])
13
14 print ('-------')
15 print ("Pearson's first coefficient of skewness: {:.2f}".
       format(a1))
16 print ("Pearson's 2nd moment of skewness: {:.2f}".format(
       a2))
17 print ("Fisher-Pearson's coefficient of skewness: {:.2f}"
       .format(g1))
18 print ('-------')
19
20 '''
21 Output:
22 -------
23 Pearson's first coefficient of skewness: 0.74
24 Pearson's 2nd moment of skewness: 0.66
25 Fisher-Pearson's coefficient of skewness: 1.26
26 -------
27 '''
```

Listing 5.8 Measuring the skewness of a data set

5.5 Descriptive Statistics in Pandas

As reported in the official pandas documentation, the command *describe()* "generates descriptive statistics that summarize the central tendency, dispersion and shape of a data set's distribution, excluding NaN (i.e., Not a Number) values." (see code Listing 5.9)

```
1  import pandas as pd
2
3  my_dataset = pd.read_excel('Smith_glass_post_NYT_data.
       xlsx', sheet_name='Supp_traces')
4
5  statistics = my_dataset[['Ba','Sr','Zr','La']].describe()
6
7  print(statistics)
8
9  '''
10 Output:
11               Ba           Sr           Zr           La
12 count    370.000000   369.000000   370.000000   370.000000
13 mean     789.733259   516.422115   365.377397    74.861088
14 std      523.974960   241.922439   118.409962    18.256772
```

15	min	0.000000	9.541958	185.416567	45.323289
16	25%	297.402777	319.667551	274.660242	61.745228
17	50%	768.562055	490.111131	339.412064	71.642167
18	75%	1278.422645	728.726116	438.847368	83.670805
19	max	2028.221963	1056.132069	920.174406	169.550008
20	'''				

Listing 5.9 Computing descriptive statistics in pandas

5.6   Box Plots

A box plot (or boxplot) uses the inter-quartile distance to describe groups of numerical data. Also, lines extending from the boxes (i.e., "whiskers") indicate the variability outside the upper and lower quartiles. The outliers are sometime plotted as individual symbols. In detail, the bottom and top of a box always represent the first and third quartiles, respectively. A line is always drawn inside the box to represents the second quartile (i.e., the median). With matplotlib, the default whisker length is 1.5 multiplied by the inter-quartile distance. Any data not included between the whiskers is considered an outlier. Using the matplotlib library, a box plot can be defined as shown in code Listing 5.10 and Fig. 5.8. Note that *dict()* defines a dictionary (see Chapter 2). Also, code Listing 5.11 and Fig. 5.9 highlight how to make box plots using the seaborn library.

```
 1  import pandas as pd
 2  import matplotlib.pyplot as plt
 3
 4  my_dataset = pd.read_excel('Smith_glass_post_NYT_data.
       xlsx', sheet_name='Supp_traces')
 5
 6  fig, ax = plt.subplots()
 7  my_flierprops = dict(markerfacecolor='#f8e9a1',
       markeredgecolor='#24305e', marker='o')
 8  my_medianprops = dict(color='#f76c6c', linewidth = 2)
 9  my_boxprops = dict(facecolor='#a8d0e6', edgecolor='#24305
       e')
10  ax.boxplot(my_dataset.Zr, patch_artist = True, notch=True
       , flierprops=my_flierprops, medianprops=my_medianprops
       , boxprops=my_boxprops)
11  ax.set_ylabel('Zr [ppm]')
12  ax.set_xticks([1])
13  ax.set_xticklabels(['all Epochs'])
14  plt.show()
```

Listing 5.10 Making a box plot using matplotlib

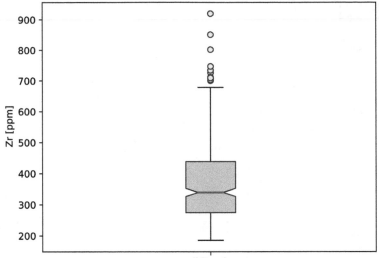

Fig. 5.8 Result of code Listing 5.10

```
1  import pandas as pd
2  import matplotlib.pyplot as plt
3  import seaborn as sns
4
5  my_dataset = pd.read_excel('Smith_glass_post_NYT_data.
       xlsx', sheet_name='Supp_traces')
6
7  fig, ax = plt.subplots()
8  ax = sns.boxplot(x="Epoch", y="Zr", data=my_dataset,
       palette="Set3")
```

Listing 5.11 Making box plots using seaborn

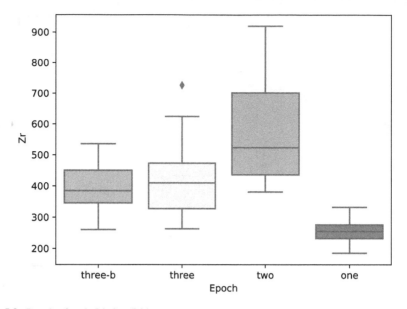

Fig. 5.9 Result of code Listing 5.11

Chapter 6
Descriptive Statistics 2: Bivariate Analysis

6.1 Covariance and Correlation

This chapter investigates how to capture relationships between two variables, which is the field of bivariate statistics. To begin, consider Fig. 6.1, which is generated by code Listing 6.1. As you can see, the two diagrams differ in Fig. 6.1. From the previous chapter, we know how to describe each variable (i.e., La, Ce, Sc, and U) appearing in Fig. 6.1 using indexes of location (e.g., the arithmetic mean), dispersion (e.g., the standard deviation), and shape (e.g., the skewness).

However, these indexes, although useful to describe a single variable, are unable to capture relationships between variables. For example, the diagram La versus Ce clearly shows an increase in Ce as La increases and vice-versa. Mathematicians describe this by using the concepts of covariance and correlation. Conversely, it is not possible to define any simple relation between Sc and U.

Definition The covariance of two sets of univariate samples x and y derived from two random variables X and Y is a measure of their joint variability, or their degree of correlation [23, 31]:

$$Cov_{xy} = \frac{\sum_{i=1}^{n}(y_i - \bar{y})(x_i - \bar{x})}{n - 1}.$$

(6.1)

$Cov_{xy} > 0$ indicates a positive relationship between Y and X. In contrast, if $Cov_{xy} < 0$, the relationship is negative [23, 31]. If X and Y are statistically independent, then $Cov_{xy} = 0$. Note that, although statistically independent variables are always uncorrelated, the converse is not necessarily true.

© The Author(s), under exclusive license to Springer Nature Switzerland AG 2021
M. Petrelli, *Introduction to Python in Earth Science Data Analysis*,
Springer Textbooks in Earth Sciences, Geography and Environment,
https://doi.org/10.1007/978-3-030-78055-5_6

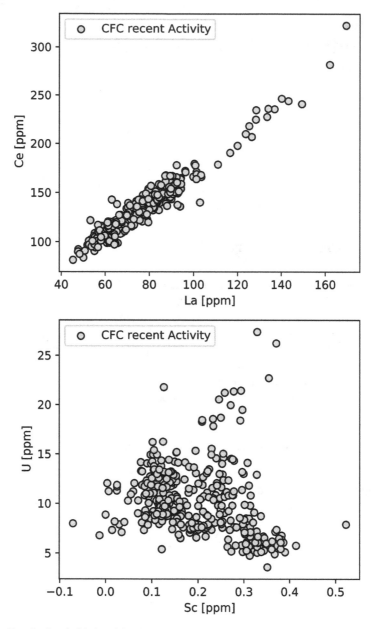

Fig. 6.1 Result of code Listing 6.1

```
1  import pandas as pd
2  import matplotlib.pyplot as plt
3
4  my_dataset = pd.read_excel('
       Smith_glass_post_NYT_data.xlsx', sheet_name='
       Supp_traces')
5
6  fig = plt.figure()
7  ax1 = fig.add_subplot(2,1,1)
8  ax1.scatter(my_dataset.La, my_dataset.Ce, marker='o'
       , edgecolor='k', color='#c7ddf4', label='CFC
       recent Activity')
9  ax1.set_xlabel('La [ppm]')
10 ax1.set_ylabel('Ce [ppm]')
11 ax1.legend()
12
13 ax2 = fig.add_subplot(2,1,2)
14 ax2.scatter(my_dataset.Sc, my_dataset.U, marker='o',
       edgecolor='k', color='#c7ddf4', label='CFC
       recent Activity')
15 ax2.set_xlabel('Sc [ppm]')
16 ax2.set_ylabel('U [ppm]')
17 ax2.legend()
```

Listing 6.1 Linear relation between two variables

Note that the covariance depends on the magnitudes of the two variables inspected. Consequently, it does not tell us much about the strength of such a relationship [23, 31]. The normalized version of the covariance (i.e., the correlation coefficient) allows us to overcome this limitation. The correlation coefficient ranges from -1 to 1 and shows, by its magnitude, the strength of the linear relation.

Equation (6.2) defines the correlation coefficient r_{xy} for two joined univariate sets of data, X and Y, characterized by a covariance Cov_{xy} and standard deviations σ_{sx} and σ_{sy}, respectively [23, 31]:

$$r_{xy} = \frac{Cov_{xy}}{\sigma_{sx}\sigma_{sy}} = \frac{\sum_{i=1}^{n}(y_i - \bar{y})(x_i - \bar{x})}{\sqrt{\sum_{i=1}^{n}(y_i - \bar{y})^2 \sum_{i=1}^{n}(x_i - \bar{x})^2}}. \tag{6.2}$$

By definition, r_{xy} is scale invariant, which means it does not depend on the magnitude of the values considered. Also, r_{xy} satisfies the following relation [23, 31]:

$$-1 \leq r_{xy} \leq 1. \tag{6.3}$$

With a pandas DataFrame, the covariance and the correlation can be readily computed by using the *cov()* and *corr()* functions. These functions calculate the covariance and the correlation matrices for a DataFrame, respectively. A covariance matrix is a table showing the covariances Cov_{xy} between variables in the DataFrame. Each cell in the table shows the covariance between two variables. The correlation matrix follows the same logic as the covariance table, but gives the correlation coefficients. In the latter, the diagonal cells all contain unity, which corresponds to the self-correlation coefficient.

Code Listing 6.2 and Fig. 6.2 report the computation and subsequent representation of the covariance and the correlation matrices for the elements reported in Fig. 6.1 (i.e., Ce, La, U, and Sc).

```
1  import pandas as pd
2  import matplotlib.pyplot as plt
3  import seaborn as sns
4
5  my_dataset = pd.read_excel('
       Smith_glass_post_NYT_data.xlsx', sheet_name='
       Supp_traces')
6
7  my_sub_dataset = my_dataset[['Ce','La','U','Sc']]
8
9  cov = my_sub_dataset.cov()
10 cor = my_sub_dataset.corr()
11
12 fig = plt.figure(figsize=(11,5))
13
14 ax1 = fig.add_subplot(1,2,1)
15 ax1.set_title('Covariance Matrix')
16 sns.heatmap(cov, annot=True, cmap='cividis', ax=
       ax1)
17
18 ax2 = fig.add_subplot(1,2,2)
19 ax2.set_title('Correlation Matrix')
20 sns.heatmap(cor, annot=True, vmin= -1, vmax=1,
       cmap='coolwarm', ax=ax2)
21
22 fig.tight_layout()
```

Listing 6.2 Estimating the covariance and the correlation matrix

Note that a value of r_{xy} close to zero only means that X and Y are not linearly related, not excluding other relationships [23, 31].

To evaluate nonlinear relationships, other parameters should be used. For example, the Spearman rank-order correlation coefficient is a non-parametric measure of the monotonicity of the relationship between two data sets. As with Pearson's correlation coefficient, the Spearman rank-order correlation coefficient ranges from -1 to $+1$, with 0 implying no correlation. Correlations of -1 or $+1$ imply an exact monotonic relationship. A positive correlation implies that Y increases as X increases. Conversely, a negative correlations implies that Y decreases as X increases. In Python, the function *scipy.stats.spearmanr()* calculates the Spearman correlation coefficient together with the associated confidence (i.e., the p-value).

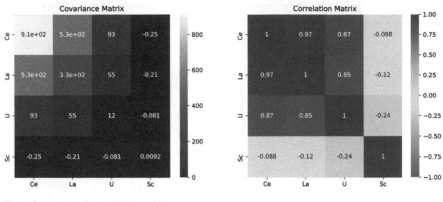

Fig. 6.2 Result of code Listing 6.2

6.2 Simple Linear Regression

Considering a response variable Y and a predictor X, we can define a linear model by using [23, 31]

$$Y = \beta_0 + \beta_1 X + \epsilon, \tag{6.4}$$

where β_0 and β_1 are coefficients respectively called the intercept (i.e., the predicted value of Y at $X = 0$) and slope (i.e., the change in Y per unit change in X). The quantity ϵ is the residual error [23, 31]. Using the least-squares method (i.e., minimizing the sum of squares of the vertical distances from each point to Eq. (6.4), β_1 and β_0 are estimated by using Eqs. (6.5) and (6.6), respectively:

$$\beta_1 = \frac{\sum_{i=1}^{n}(y_i - \bar{y})(x_i - \bar{x})}{\sum_{i=1}^{n}(x_i - \bar{x})^2} = \frac{Cov_{xy}}{\sigma_{sx}^2} = r_{xy}\frac{\sigma_{sy}}{\sigma_{sx}}, \tag{6.5}$$

$$\beta_0 = \bar{y} - \beta_1\bar{x}. \tag{6.6}$$

The square of the correlation coefficient, r_{xy}^2, with $0 \leq r_{xy}^2 \leq 1$, is typically used to make a preliminary estimate of the quality of the given regression model.

A more exhaustive evaluation of the model requires a detailed analysis of the errors, (i.e., an error analysis), which is discussed in Chap. 10.

Python contains numerous implementations of the least-squares method for first-order linear regression. Examples are the the *linregress()* function (code Listing 6.3 and Fig. 6.3) in the statistical module of Scipy and the linear regression module in statsmodels.[1]

---

[1] https://www.statsmodels.org.

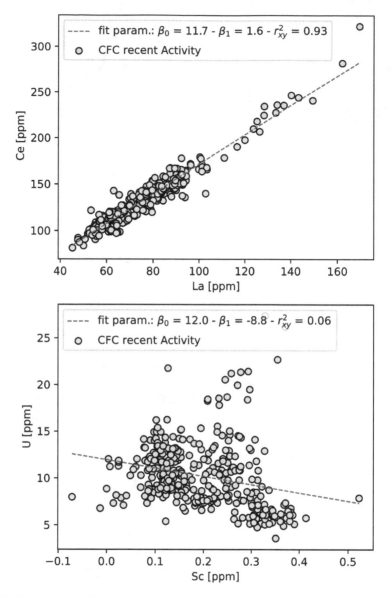

Fig. 6.3 Result of code Listing 6.3

```
 1 import pandas as pd
 2 import scipy.stats as st
 3 import numpy as np
 4 import matplotlib.pyplot as plt
 5
 6 my_dataset = pd.read_excel('
       Smith_glass_post_NYT_data.xlsx', sheet_name='
       Supp_traces')
 7
 8 fig = plt.figure()
 9 ax1= fig.add_subplot(2,1,1)
10 ax1.scatter(my_dataset.La, my_dataset.Ce, marker=
       'o', edgecolor='k', color='#c7ddf4', label='
       CFC recent Activity')
11 b1, b0, rho_value, p_value, std_err = st.
       linregress(my_dataset.La, my_dataset.Ce)
12 x = np.linspace(my_dataset.La.min(),my_dataset.La
       .max())
13 y = b0 + b1*x
14 ax1.plot(x, y, linewidth=1, color='#ff464a',
       linestyle='--', label=r"fit param.: $\beta_0$
       = " + '{:.1f}'.format(b0) + r" - $\beta_1$ = "
       + '{:.1f}'.format(b1) + r" - $r_{xy}^{2}$ =
       " + '{:.2f}'.format(rho_value**2))
15 ax1.set_xlabel('La [ppm]')
16 ax1.set_ylabel('Ce [ppm]')
17 ax1.legend(loc= 'upper left')
18
19 ax2 = fig.add_subplot(2,1,2)
20 ax2.scatter(my_dataset.Sc, my_dataset.U, marker='
       o', edgecolor='k', color='#c7ddf4', label='CFC
       recent Activity')
21 b1, b0, rho_value, p_value, std_err = st.
       linregress(my_dataset.Sc, my_dataset.U)
22 x = np.linspace(my_dataset.Sc.min(),my_dataset.Sc
       .max())
23 y = b0 + b1*x
24 ax2.plot(x, y, linewidth=1, color='#ff464a',
       linestyle='--', label=r"fit param.: $\beta_0$
       = " + '{:.1f}'.format(b0) + r" - $\beta_1$ = "
       + '{:.1f}'.format(b1) + r" - $r_{xy}^{2}$ =
       " + '{:.2f}'.format(rho_value**2))
25 ax2.set_xlabel('Sc [ppm]')
26 ax2.set_ylabel('U [ppm]')
27 ax2.legend(loc= 'upper left')
```

Listing 6.3 Least-squares linear regression applied to the data of Fig. 6.1

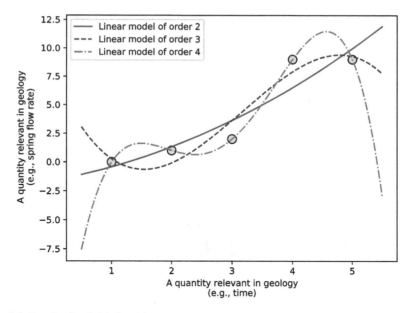

Fig. 6.4 Result of code Listing 6.4

6.3 Polynomial Regression

The linear model defined in Eq. (6.4) is easily generalized to a polynomial of degree n:

$$Y = \beta_0 + \beta_1 X + \beta_2 X^2 + \beta_3 X^3 + \cdots + \beta_n X^n + \epsilon. \tag{6.7}$$

If $n > 1$, the function $Y(X)$ is nonlinear but the regression model remains linear because the regression parameters $\beta_0, \beta_1, \beta_2, \ldots, \beta_n$ enter Eq. (6.7) as linear terms [23, 31].

For example, suppose you collected a geological quantity (e.g., the flow of spring water) at selected time intervals and you wish to fit your data with second-, third-, and fourth-order polynomial models. Code Listing 6.4 shows how to do this in Python by using the *numpy.polyfit()* function (see code Listing 6.4 and Fig. 6.4).

```
1 import numpy as np
2 import matplotlib.pyplot as plt
3
4 x = np.arange(1,6)
5 y = np.array([0,1,2,9,9])
6
7 fig, ax = plt.subplots()
8 ax.scatter(x, y, marker = 'o', s = 100, color = '#
      c7ddf4', edgecolor = 'k')
9
```

```
10  orders = np.array([2,3,4])
11  colors =['#ff464a','#342a77','#4881e9']
12  linestiles = ['-','--','-.']
13
14  for order, color, linestile in zip(orders, colors,
         linestiles):
15      betas = np.polyfit(x, y, order)
16      func = np.poly1d(betas)
17      x1 = np.linspace(0.5,5.5, 1000)
18      y1 = func(x1)
19      ax.plot(x1, y1, color=color, linestyle=linestile
         , label="Linear model of order " + str(order))
20
21  ax.legend()
22  ax.set_xlabel('A quantity relevant in geology\n(e.g
         ., time)')
23  ax.set_ylabel('A quantity relevant in geology\n(e.g
         ., spring flow rate)')
24  fig.tight_layout()
```

Listing 6.4 Polynomial regression of order n

6.4 Nonlinear Regression

In regression analysis, the linear and nonlinear terms do not describe the relationship between Y and X but instead refer to regression parameters that enter the equation linearly or nonlinearly [23, 31]. For example, regression models for Eqs. (6.4) and (6.7) are both linear. Also, the regression model for Eq. (6.8) is linear too. In detail, both Eqs. (6.7) and (6.8) can be transformed into a linear form [23] [i.e., Eqs. (6.9) and (6.10)].

$$Y = \beta_0 + \beta_1 \log(X) + \epsilon. \tag{6.8}$$

For example, we can set $X^2 = X_2, X^3 = X_3, \ldots, X^n = X_n$ in Eq. (6.7) and $X_1 = \log(X)$ in Eq. (6.8). The resulting equations are both linear:

$$Y = \beta_0 + \beta_1 X + \beta_2 X_2 + \beta_3 X_3 + \cdots + \beta_n X_n + \epsilon, \tag{6.9}$$
$$Y = \beta_0 + \beta_1 X_1 + \epsilon. \tag{6.10}$$

In general, all regression parameters enter the equation linearly in linear models, possibly after transforming the data [23]. In contrast, in nonlinear models, the relationship between Y and some of the predictors is nonlinear, or some of the parameters appear nonlinearly, but no transformation is possible to make the parameters appear linearly [23]. Table 6.1 provides a checklist, modified from [33], to determine whether linear regression is appropriate for your data set.

Table 6.1 Is linear regression appropriate for your geological data set? Modified from [33]

Question	Discussion
Are X and Y related by a straight line?	For many geological applications, the relationship between X and Y is nonlinear, making linear regression inappropriate. You should either transform the data or perform nonlinear curve fitting
Is the scattering of data around the line normally distributed?	Linear regression analysis assumes that the scatter is Gaussian
Is variability the same everywhere?	Linear regression assumes that the scattering around the best-fit line has the same standard deviation all along the curve. The assumption is violated if the points with high or low X values tend to be farther from the best-fit line (i.e., homoscedasticity)
Are the X values known with precision?	The least-squares linear regression model assumes that X values are exactly correct, which means that X is very small compared with the variability in Y, and that experimental error or geological variability only affects the Y values
Are the data points independent?	Whether a point is above or below the line is a matter of chance and does not influence whether another point is above or below the line
Are the X and Y values intertwined?	If X is used to calculate Y (or vice-versa), then linear regression calculations are invalid

An example of nonlinear regression in petrology is given by the application of the crystal-lattice-strain model [20] to interpret experimental data. In detail, this model provides a conceptual framework for quantifying partition coefficients D_i in magmatic systems [20, 29], where D_i is given by

$$D_i = D_0 \exp \left\{ \frac{-4\pi E N_A \left[\frac{r_0}{2}(r_i - r_0)^2 + \frac{1}{3}(r_i - r_0)^3 \right]}{RT} \right\}, \qquad (6.11)$$

where T is the temperature, r_i is the radius the trace element i belonging to an isovalent set of elements, r_0 is the radius of the ideal element that minimally strains the crystal lattice (i.e., characterized by the largest D_i), D_0 is the partition coefficient of the ideal element characterized by radius r_0, E is the apparent Young's modulus of the site, and N_A and R are Avogadro's number and the universal gas constant, respectively [20, 29]. Equation (6.11) plots near-parabolically with ionic radius if you graph $\log_{10}(D_i)$ versus r_i [21].

Typically, r_0, D_0, and E are estimated by fitting by non-linear regression Eq. (6.11) to the experimentally determined D_i.

```
 1 import numpy as np
 2 import matplotlib.pyplot as plt
 3 from scipy.optimize import curve_fit
 4
 5 def func(r, r0, D0, E):
 6     R=8.314462618
 7     scale = 1e-21 # r in Angstrom (r^3 -> 10^-30 m)
    , E is GPa (10^9 Pa)
 8     T = 800 + 273.15
 9     Na=6.02e23
10     return D0*np.exp((-4*np.pi*E*Na*((r0/2)*(r-r0)
    **2+(1/3)*(r-r0)**3)*scale)/(R*T))
11
12 def add_elements(ax):
13     # to plot the name of the elements on the
    diagram
14     names = ['La', 'Ce', 'Nd', 'Sm', 'Eu', 'Gd', '
    Dy', 'Er', 'Yb', 'Lu', 'Y', 'Sc']
15     annotate_xs = np.array([1.172 + 0.01, 1.15 +
    0.01, 1.123 + 0.01, 1.098 - 0.031, 1.087 -
    0.028, 1.078 - 0.04, 1.052 + 0.005, 1.03 +
    0.02, 1.008 + 0.01, 1.001 - 0.015, 1.04 -0.02,
    0.885 - 0.03])
16     annotate_ys = np.array([0.468 + 0.1, 1.050 +
    0.2, 10.305 + 3, 31.283 - 13,  45.634 -17,
    74.633- 30, 229.279 + 80, 485.500, 583.828
    +200, 460.404 -220, 172.844 -70, 141.630])
17
18     for name, annotate_x, annotate_y in zip(names,
    annotate_xs, annotate_ys):
19         ax.annotate(name, (annotate_x, annotate_y))
20
21 Di = np.array([0.468, 1.050, 10.305,  31.283,
    45.634, 74.633, 229.279, 485.500, 583.828,
    460.404, 172.844, 141.630])
22 I_r = np.array([1.172, 1.15, 1.123, 1.098, 1.087,
    1.078, 1.052, 1.03, 1.008, 1.001, 1.04, 0.885])
23
24 fig = plt.figure(figsize=(9,5))
25
26 # Trust Region Reflective algorithm
27 ax1 = fig.add_subplot(1,2,1)
28 ax1.set_title("Trust Region Reflective algorithm")
29 ax1.scatter(I_r, Di, s=80, color='#c7ddf4',
    edgecolors='k', label='4 GPa - 1073 K, Kessel
    et al., 2005')
30
31 popt1, pcov1 = curve_fit(func, I_r, Di, method='trf
    ', bounds=([0.8,0,0],[1.3,1000,1000]))
32
33 x1 = np.linspace(0.85,1.2,1000)
34 y1 = func(x1,popt1[0],popt1[1], popt1[2])
```

```
35 ax1.plot(x1,y1, color='#ff464a', linewidth=2,
      linestyle ='--', label=r'$r_0$ = ' + '{:.3f}'.
      format(popt1[0]) + r', $D_0$ = ' + '{:.0f}'.
      format(popt1[1]) + ', E = ' + '{:.0f}'.format(
      popt1[2]))
36 add_elements(ax = ax1)
37 ax1.set_yscale('log')
38 ax1.set_xlabel(r'Ionic Radius ($\AA$)')
39 ax1.set_ylabel(r'$D_i$')
40 ax1.set_ylim(0.005,3000)
41 ax1.legend()
42
43 # Levenberg-Marquardt algorithm
44 ax2 = fig.add_subplot(1,2,2)
45 ax2.set_title("Levenberg-Marquardt algorithm")
46 ax2.scatter(I_r, Di, s=80, color='#c7ddf4',
      edgecolors='k', label='4 GPa - 1073 K, Kessel
      et al., 2005')
47
48 popt2, pcov2 = curve_fit(func, I_r, Di, method='lm'
      , p0=(1.1,100,100))
49
50 x2 = np.linspace(0.85,1.2,1000)
51 y2 = func(x2,popt2[0],popt2[1], popt2[2])
52 ax2.plot(x2,y2, color='#4881e9', linewidth=2,
      linestyle ='--', label=r'$r_0$ = ' + '{:.3f}'.
      format(popt2[0]) + r', $D_0$ = ' + '{:.0f}'.
      format(popt2[1]) + ', E = ' + '{:.0f}'.format(
      popt2[2]))
53 add_elements(ax = ax2)
54 ax2.set_yscale('log')
55 ax2.set_xlabel(r'Ionic Radius ($\AA$)')
56 ax2.set_ylabel(r'$D_i$')
57 ax2.set_ylim(0.005,3000)
58 ax2.legend()
59
60 fig.tight_layout()
```

Listing 6.5 Least-squares nonlinear regression to extract r_0, D_0, and E from an experimental set of D_i in the framework of the crystal-lattice-strain model [20]

In Python, the function *scipy.optimize.curve_fit()* applies the nonlinear least-squares method to fit a function to data and can be used to extract r_0, D_0, and E from experimental results for D_i. In detail, *curve_fit()* is based on three algorithms: the trust region reflective algorithm [22], the dogleg algorithm with rectangular trust regions [38], and the Levenberg-Marquardt algorithm [32]. A detailed description of the algorithms governing the nonlinear regression is beyond the scope of this book but is available in more specialized books (see, e.g., [36]). As an example, code Listing 6.5 replicates the results reported in Fig. 2 of [29].

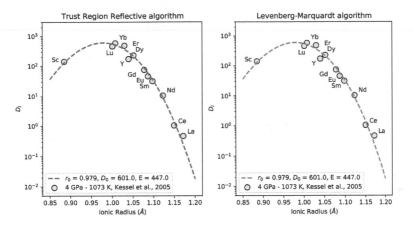

Fig. 6.5 Result of code Listing 6.5

Figure 6.5 shows the best fit of Eq. (6.11) using the trust region reflective algorithm [22] with bounds for r_0, D_0, and E (line 31 of code Listing 6.5) and the Levenberg-Marquardt algorithm [32] with an initial guess of r_0, D_0, and E (see p0, line 48 of code Listing 6.5). The two algorithms return the same best-fit parameters (Fig. 6.5).

Part III
Integrals and Differential Equations in Geology

Chapter 7
Numerical Integration

7.1 Definite Integrals

From the operational point of view, integration mainly involves problems of two different classes [57]. The first class of problems are indefinite integrals and involve functions that we must find given their derivative [57]. The second class of problems are definite integrals, which consist of summing up a large number of extremely small quantities to find areas, volumes, centers of gravity, etc. [57].

For most geological applications, problems involving integrals can be reduced to definite integrals.

Informal definition: Given a function f of a real variable x, the definite integral (S) of $f(x)$ over an interval of real numbers $[a, b]$ is the area bounded by $f(x)$, the x axis, and the vertical lines at $x = a$ and $x = b$ (Fig. 7.1).

Note that the regions above and below the x axis enter the sum with a positive and negative sign, respectively (Fig. 7.2).

7.2 Basic Properties of Integrals

Definite integrals have some interesting properties that are often useful for solving complex problems by reducing them to simpler problems. Three such properties are listed below:

Additive properties:

© The Author(s), under exclusive license to Springer Nature Switzerland AG 2021
M. Petrelli, *Introduction to Python in Earth Science Data Analysis*,
Springer Textbooks in Earth Sciences, Geography and Environment,
https://doi.org/10.1007/978-3-030-78055-5_7

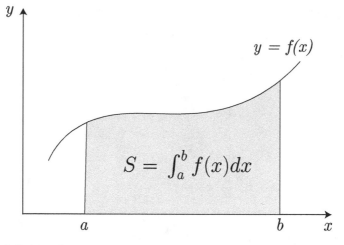

Fig. 7.1 Definite integral

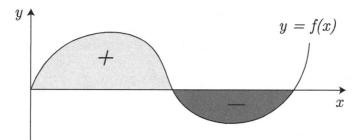

Fig. 7.2 Sign of definite integrals

$$\int_a^b f(x)dx + \int_b^c f(x)dx = \int_a^c f(x)dx, \tag{7.1}$$

$$\int_a^a f(x)dx = 0, \tag{7.2}$$

$$\int_a^b f(x)dx = -\int_b^a f(x)dx. \tag{7.3}$$

Scaling by a constant:

$$\int_a^b cf(x)dx = c\int_a^b f(x)dx. \tag{7.4}$$

Integral of a sum:

$$\int_a^b [f(x) + g(x)]dx = \int_a^b f(x)dx + \int_a^b g(x)dx. \tag{7.5}$$

7.3 Analytical and Numerical Solutions of Definite Integrals

In general, analytical methods give exact solutions, but sometimes such solutions are impossible to achieve. In contrast, numerical methods give approximate solutions with allowable tolerances (i.e., an error characterized by a known confidence limit). Also, numerical methods are our only recourse when a function is only empirically estimated at discrete points, as in most cases dealing with geological sampling (e.g., volatile fluxes at volcanic areas).

A detailed description of the analytical solutions of definite integrals is beyond the scope of this book, so in the following I simply provide without proof the definition from the Fundamental Theorem of Calculus and give a few simple examples based on the symbolic approach in Python. Numerical methods, however, are discussed in detail, mainly focusing on algorithms that solve definite integrals even when $f(x)$ is not mathematically defined [i.e., when we only know a limited number of fixed values of $f(x)$], as in the case of sampling in many geological fields.

7.4 Fundamental Theorem of Calculus and Analytical Solutions

Fundamental Theorem of Calculus

The Fundamental Theorem of Calculus formulates an analytical link between differentiation and integration. The theorem consists of two parts, the first of which establishes the relationship between differentiation and integration [57, 59].

Part 1. If $F(x)$ is continuous over an interval $[a, b]$ and the function $F(x)$ is defined by

$$F(x) = \int_a^x f(t)dt, \tag{7.6}$$

then $F'(x) = f(x)$ over $[a, b]$, and we define $F(x)$ as the "antiderivative" of $f(x)$.

The second part of the theorem affirms that, if we can determine an antiderivative for the integrand, then we can evaluate the definite integral by evaluating the antiderivative at the extreme points of the interval and subtracting.

Part 2. If $f(x)$ is continuous over the interval $[a, b]$ and $F(x)$ is the antiderivative of $f(x)$, then

$$\int_a^b f(x)dx = F(b) - F(a). \tag{7.7}$$

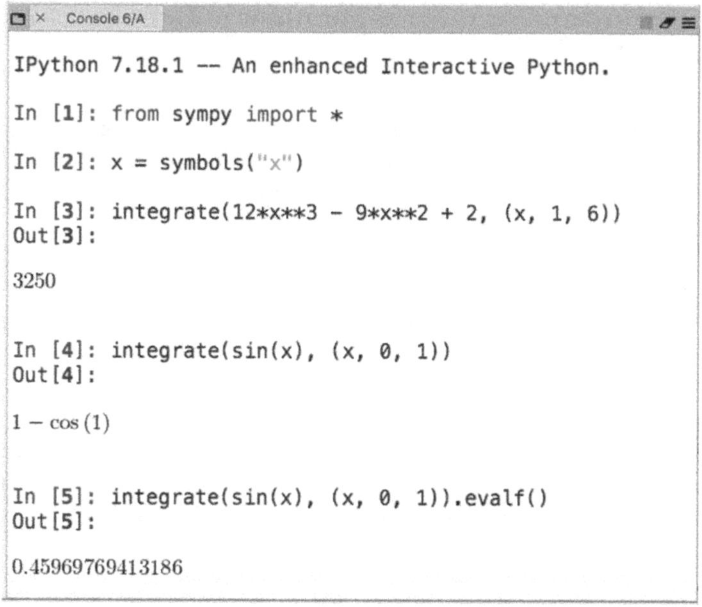

```
  □  ×    Console 6/A                                              ■ ◢ ≡

 IPython 7.18.1 -- An enhanced Interactive Python.

 In [1]: from sympy import *

 In [2]: x = symbols("x")

 In [3]: integrate(12*x**3 - 9*x**2 + 2, (x, 1, 6))
 Out[3]:

 3250


 In [4]: integrate(sin(x), (x, 0, 1))
 Out[4]:

 1 - cos (1)


 In [5]: integrate(sin(x), (x, 0, 1)).evalf()
 Out[5]:

 0.45969769413186
```

Fig. 7.3 Symbolic integration using SimPy

Analytical Solutions: The Symbolic Approach in Python

Symbolic computation symbolically manipulates and solves mathematical expressions [12]. In symbolic computation, mathematical objects are represented exactly and not approximately, as in the case of numerical solutions [12]. Also, mathematical expressions with unevaluated variables are left in the symbolic form [12]. In Python, the SimPy package uses the symbolic approach to simplify expressions, compute derivatives, integrals, and limits, solve equations, work with matrices, etc. [12].

As a simple example consider Fig. 7.3, which shows how to use SimPy to analytically solve the following two definite integrals:

$$\int_1^6 12x^3 - 9x^2 + 2dx = \left[3x^4 - 3x^3 + 2x\right]_1^6 = (3252 - 2) = 3250, \quad (7.8)$$

$$\int_0^1 \sin(x)dx = [-\cos(x)]_0^1 = 1 - \cos(1) \simeq 0.46. \quad (7.9)$$

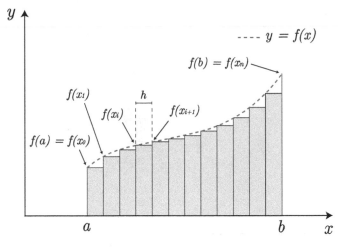

Fig. 7.4 Example of the "left-rectangle" approximation to solve definite integrals

7.5 Numerical Solutions of Definite Integrals

Rectangle method

The simplest method to numerically approximate the solution of a definite integral is to divide the area of interest into many rectangles of equal width and variable height, and then sum up the area of each rectangle to obtain the area under the curve (i.e., the definite integral; see Fig. 7.4):

$$\int_a^b f(x)dx \approx h \sum_{i=0}^{n-1} f(x_i),$$ (7.10)

where n is the number of rectangles, $x_0 = a$, $x_n = b$, and

$$h = \frac{b-a}{n}.$$ (7.11)

The procedure reported in Eq. (7.10) and Fig. 7.4 is the so-called "left-rectangle" approximation. Additional options include the "right-" [Eq. (7.12)] and "midpoint-rectangle" [Eq. (7.13)] approximations (Fig. 7.5):

$$\int_a^b f(x)dx \approx h \sum_{i=1}^{n} f(x_i),$$ (7.12)

$$\int_a^b f(x)dx \approx h \sum_{i=0}^{n-1} \frac{(f(x_i) + f(x_{i+1}))}{2}.$$ (7.13)

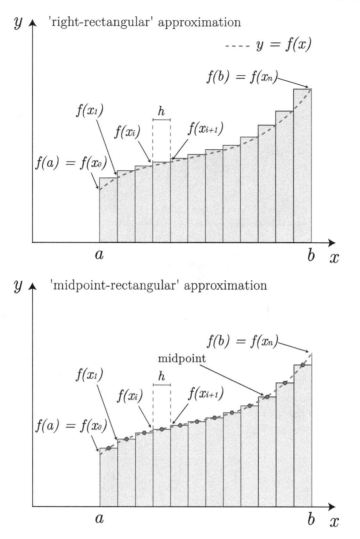

Fig. 7.5 Right- and midpoint-rectangle approximations for solving definite integrals

Filling $f(x)$ with rectangles roughly approximates the area of interest. However, the more rectangles are inserted between the boundaries a and b, the more accurate the approximation will be because the uncovered regions become smaller.

We can write a simple function in Python to implement the rectangle method (code Listing 7.1).

```
 1 import numpy as np
 2
 3 def integrate_rec(f, a, b, n):
 4     # Implementation of the rectanlge method
 5     h = (b-a)/n
 6     x = np.linspace(a, b, n+1)
 7     i = 0
 8     area = 0
 9     while i < n:
10         sup_rect =  f(x[i])*h
11         area += sup_rect
12         i += 1
13     return area
14 '''
15 We test the Rectangle method on the sine funcion
       were the definite integral in the interval [0,
       π/2] is equal to 1.
16 '''
17
18 sup_5 = integrate_rec(np.sin, 0, np.pi/2, 5)
19 sup_10 = integrate_rec(np.sin, 0, np.pi/2, 10)
20 sup_100 = integrate_rec(np.sin, 0, np.pi/2, 100)
21
22 print('Using n=5, the rectangle method returns a
       value of {:.2f}'.format(sup_5))
23 print('Using n=10, the rectangle method returns a
        value of {:.2f}'.format(sup_10))
24 print('Using n=100, the rectangle method returns
       a value of {:.2f}'.format(sup_100))
25
26 '''
27 Output:
28 Using n=5, the rectangle method returns a value
        of 0.83
29 Using n=10, the rectangle method returns a value
        of 0.92
30 Using n=100, the rectangle method returns a value
        of 0.99
31 '''
```

Listing 7.1 Rectangle rule to solve definite integrals

Trapezoid rule

The trapezoid rule is the basis for a technique similar to the rectangle method. Instead of rectangles, the trapezoid rule uses trapezoids to fill the area under $f(x)$ (Fig. 7.6). Equation (7.14) and code Listing 7.2 report the mathematical formulation of the trapezoid rule and its implementation in Python, respectively:

$$S = \int_a^b f(x)dx \approx h\left[\frac{f(x_0) + f(x_n)}{2} + \sum_{i=1}^{n-1} f(x_i)\right]. \qquad (7.14)$$

```python
1  import numpy as np
2
3  def integrate_trap(f, a, b, n):
4      # Implementation of the trapezoidal rule
5      h = (b-a)/n
6      x = np.linspace(a, b, n+1)
7      i=1
8      area = h*(f(x[0]) + f(x[n]))/2
9      while i<n:
10         sup_rect = f(x[i])*h
11         area += sup_rect
12         i += 1
13     return area
14
15  '''
16  We test the trapezoidal rule on the known sine
       funcion were the
17  definite integral in the interval [0, π/2] is
       equal to 1.
18  '''
19
20  sup_5 = integrate_trap(np.sin, 0, np.pi/2, 5)
21  sup_10 = integrate_trap(np.sin, 0, np.pi/2, 10)
22
23  print('Using n=5, the trapezoidal rule returns a
       value of {:.2f}'.format(sup_5))
24  print('Using n=10, the trapezoidal rule returns a
       value of {:.2f}'.format(sup_10))
25
26  '''
27  Output:
28  Using n=5, the trapezoidal rule returns a value
       of 0.99
29  Using n=10, the trapezoidal rule returns a value
       of 1.00
30  '''
```

Listing 7.2 Trapezoid rule to solve definite integrals

Trapezoidal and composite Simpson rules using scipy

The scipy.integrate sub-package implements many techniques for solving definite integrals, including the trapezoid rule mentioned in the previous paragraph and the composite Simpson rule.

The composite Simpson rule is a technique that approximates an integral over each pair of consecutive sub-intervals using quadratic functions (Fig. 7.7). The resulting

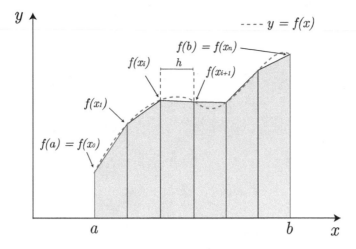

Fig. 7.6 Trapezoid rule to solve definite integrals

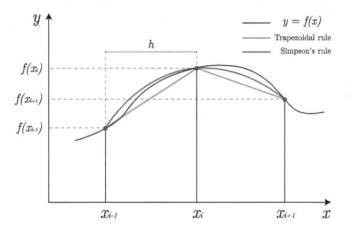

Fig. 7.7 Example of the composite Simpson's rule and comparison with the trapezoid rule

formula to calculate a definite integral is

$$S = \int_a^b f(x)dx \approx \frac{h}{3} \sum_{i=1}^{n/2} \left[f(x_{2i-2}) + 4f(x_{2i-i}) + f(x_{2i}) \right], \qquad (7.15)$$

where n is an even number that gives the number of sub-intervals of $[a, b]$, as done for the rectangle methods and the trapezoid rule.

Code Listing 7.3 uses the scipy.integrate sub-package to apply the trapezoid and composite Simpson rules to the equation $y = x^2$ (Fig. 7.8).

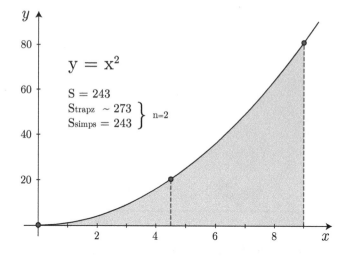

Fig. 7.8 Application of trapezoid and composite Simpson rules to the equation $y = x^2$ in the interval $[0, 9]$. The analytical result is 243. Given that $y = x^2$ is a quadratic function, it is perfectly fit by Simpson's rule

```
1  import numpy as np
2  from scipy import integrate
3
4
5  x = np.linspace(0,9, 3) # 3 divisions [x_0, x_1, x_2], n=2
6  y = x**2
7
8  sup_trapz = integrate.trapz(y,x)
9  sup_simps = integrate.simps(y,x)
10
11
12 print('Using n=2, the trapezoidal rule returns a
        value of {:.0f}'.format(sup_trapz))
13 print('Using n=2, the composite Simpson rule
        returns a value of {:.0f}'.format(sup_simps))
14
15 '''
16 Output:
17 Using n=2, the trapezoidal rule returns a value
        of 273
18 Using n=2, the composite Simpson rule returns a
        value of 243
19 '''
```

Listing 7.3 Application of trapezoid and composite Simpson rules to the equation $y = x^2$

7.6 Computing the Volume of Geological Structures

An application in geology of definite integrals is to estimate the volume of structures that cannot be approximated by simple geometries. For example, estimating volumes is one of the most basic and widely applied tasks of hydrocarbon exploration and production [58].

Qualitatively, to approximate the volume of a solid, we imagine slicing it into many parts. We then estimate the volume of each part by using quantifiable geometries (e.g., trapezoidal prisms). Finally, we sum all the volumes to make our estimate [58, 59].

Quantitatively, if the distance between two successive slice planes is infinitesimal, we can mathematically express the procedure by using a definite integral:

$$V = \int_a^b A(x)dx, \tag{7.16}$$

```
1  import numpy as np
2  from scipy import integrate
3
4  conturs_areas = np.array([194135, 136366, 79745,
       38335, 18450, 9635, 3895])
5  x = np.array([0,25,50,75,100,125,150])
6
7  vol_traps = integrate.trapz(conturs_areas, x)
8  vol_simps = integrate.simps(conturs_areas, x)
9
10 print('The trapezoidal rule returns a volume of
       {:.0f} cubic meters'.format(vol_traps))
11 print('The composite Simpson rule returns a
       volume of {:.0f} cubic meters'.format(
       vol_simps))
12
13 '''
14 Output:
15 The trapezoidal rule returns a volume of 9538650
       cubic meters
16 The composite Simpson rule returns a volume of
       9431367 cubic meters
17 '''
```

Listing 7.4 Application of Eq. (7.16) to estimate the volume of the geological structure shown in Fig. 7.9

where V is the volume of a solid extending from $x = a$ to $x = b$, and $A(x)$ is the area of the intersection of the volume with a plane parallel to the y-z plane and passing through the point $(x, 0, 0)$ (Fig. 7.9).

Code Listing 7.4 shows how to apply Eq. (7.16) to the structure shown in Fig. 7.9.

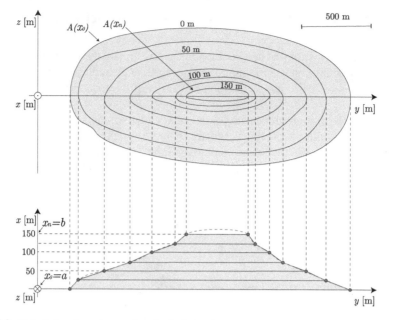

Fig. 7.9 Calculating the volume of geological structures

7.7   Computing the Lithostatic Pressure

We define the lithostatic pressure as the vertical pressure at a specific depth due to the weight of the overlying rocks. The pressure applied by a resting rock mass (assuming adiabatic compression, hydrostatic equilibrium, and spherical symmetry) under the acceleration of gravity [56] is related to the rock density by

$$p(z) = p_0 + \int_{z_1=0}^{z} \rho(z_1)g(z_1)dz_1, \tag{7.17}$$

where $p(z)$ is the pressure at depth z, p_0 is the pressure at the surface, $\rho(z_1)$ is the bulk density for the rock mass as a function of depth, and $g(z_1)$ is the acceleration due to gravity.

As a zeroth-order approximation, we assume $p_0 = 0$ and that both $\rho(z)$ and $g(z)$ are constant, which reduces Eq. (7.17) to

$$p(z) = \rho g z. \tag{7.18}$$

Code Listing 7.5 shows the implementation of Eq. (7.18) in Python.

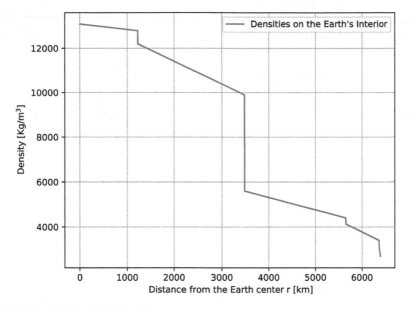

Fig. 7.10 Result of code Listing 7.6

```
1 def simple_lithopress(z,ro=2900, g=9.8):
2     pressure = z*g*ro/1e6 # return the pressure in
      MPa
3     return pressure
4
5 my_pressure = simple_lithopress(z=2000)
6 print('The pressure at 2000 meters is {0:.0f} MPa'.
      format(my_pressure))
7
8 '''
9 Output: The pressure at 2000 meters is 57 MPa
10 '''
```

Listing 7.5 Simple listing showing the implementation of Eq. (7.18) in Python

We now take on the full implementation of Eq. (7.17). Using the data from [40, 46] and assuming that ρ varies linearly between the upper and lower limits of each shell (i.e., crust, upper mantle, lower mantle, outer core, and inner core), we create an array that constitutes a first-order approximation of $\rho(z)$ (Table 7.1, code Listing 7.6, and Fig. 7.10).

Table 7.1 Earth's shells and relative densities

Layer	r from (km)	r to (km)	Thickness (km)	Bottom density (kg/m^3)	Top density (kg/m^3)
Inner core	1	1220	1220	13 100	12 800
Outer core	1221	3479	2259	12 200	9900
Lower mantle	3480	5650	2171	5600	4400
Upper mantle	5651	6370	720	4130	3400
Crust	6371	6400	30	3100	2700

```
1  import numpy as np
2  from scipy.integrate import trapz
3  import matplotlib.pyplot as plt
4
5  r = np.linspace(1,6400,6400)
6
7  def density():
8      ro_inner_core = np.linspace(13100, 12800,
         1220)
9      ro_outer_core = np.linspace(12200, 9900,
         2259)
10     ro_lower_mantle = np.linspace(5600,4400,2171)
11     ro_upper_mantle = np.linspace(4130,3400,720)
12     ro_crust = np.linspace(3100,2700,30)
13
14     ro_final = np.concatenate((ro_inner_core,
         ro_outer_core, ro_lower_mantle,
         ro_upper_mantle, ro_crust))
15
16     return ro_final
17
18 ro = density()
19
20 fig, ax = plt.subplots()
21 ax.plot(r,ro, label="Densities on the Earth's
      Interior")
22 ax.set_ylabel(r"Density [Kg/m$^3$]")
23 ax.set_xlabel("Distance from the Earth center r [
      km]")
24 ax.legend()
25 ax.grid()
```

Listing 7.6 Defining the densities in Earth's interior

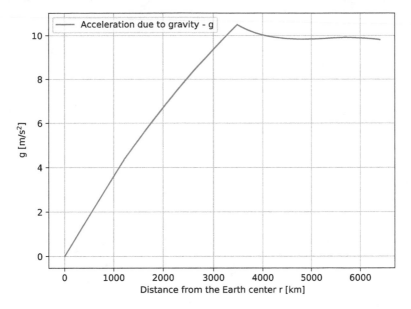

Fig. 7.11 Result of code Listing 7.7

To simplify the presentation of the data, we define a new variable r (the radial distance from the Earth's center) as $r = R - z$, where $R \approx 6400$ km is the Earth's radius and the Earth is approximated as a sphere.

The acceleration $g(r)$ at a distance r from Earth's center [56] is estimated by using

$$g(r) = \frac{4\pi G}{r^2} \int_{r_1=0}^{r} \rho(r_1)r_1^2 dr_1, \tag{7.19}$$

where $G = (6.67408 \pm 0.0031) \times 10^{-11} \text{ m}^3 \text{ kg}^{-1} \text{ s}^{-2}$ is the "universal gravitational constant" (see code Listing 7.7 and Fig. 7.11). Finally, code Listing 7.8 and Fig. 7.12 report the full implementation of Eq. (7.17), from the Earth's surface (z=0 km) to the Earth's center (z≈6400 km)

```
 1 def gravity(r):
 2
 3     g = np.zeros(len(r))
 4     Gr = 6.67408e-11
 5     r = r * 1000 # from Km to m
 6
 7     for i in range(1,len(r)):
 8
 9         r1 = r[0:i]
10         ro1 = ro[0:i]
11         r2 = r1[i-1]
12
13         y = ro1*r1**2
14         y_int = trapz(y,r1)
15
16         g1 = ((4 * np.pi*Gr)/(r2**2)) * y_int
17         g[i] = g1
18
19     return g
20
21 g = gravity(r)
22
23 fig, ax = plt.subplots()
24 ax.plot(r,g)
25 ax.grid()
26 ax.set_ylabel(r'g [m/s$^2]$')
27 ax.set_xlabel('Distance from the Earth center r [km]
       ')
```

Listing 7.7 Computing the acceleration due to gravity in Earth's interior

```
 1 def pressure(r, ro, g):
 2
 3     p = np.zeros(len(r))
 4     r = r *1000
 5
 6     for i in range(0,len(r)):
 7         r1 = r[i:len(r)]
 8         ro1 = ro[i:len(r)]
 9         g1 = g[i:len(r)]
10         y = ro1*g1
11         p1 = trapz(y,r1)
12         p[i] = p1
13     return p
14
15 p = pressure(r,ro,g)/1e9 # expressed in GPa
16 z = np.linspace(6400, 1, 6400)
17
18 fig, ax = plt.subplots()
19 ax.plot(z,p)
20 ax.grid()
21 ax.set_ylabel('P [GPa]')
22 ax.set_xlabel('Depth z from the Earth Surface [km]')
```

Listing 7.8 Computing the pressure in Earth's interior

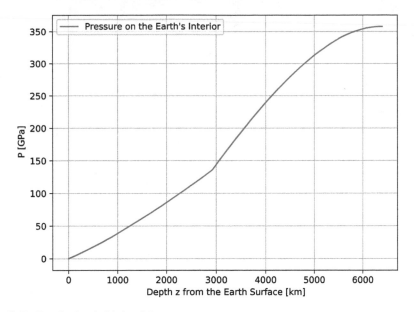

Fig. 7.12 Result of code Listing 7.8

Chapter 8
Differential Equations

8.1 Introduction

Definition A differential equation is an equation that relates one or more functions and their derivatives [60].

Qualitatively, a differential equation describes the rate at which one variable changes with respect to another. Examples include the rate of change in the number of atoms of a radioactive material over time or the rate of change in magma temperature during cooling [42]. An equation is called an ordinary differential equation (ODE) if it contains ordinary derivatives only [60]. In other words, an ODE depends on a single independent variable. To clarify the concept, ODEs have one independent variable (e.g., t), one dependent variable [e.g., $y = N(t)$], and the derivative of the dependent variable with respect to the independent variable (e.g., dN/dt). Everything else apart from the independent variable, the dependent variable, and the derivatives are constants [49]. The law of radioactive decay (i.e., the change in radioactive material per unit time) is an example of an ODE:

$$\frac{dN}{dt} = -\lambda N(t), \tag{8.1}$$

where $N(t)$ is the number of radioactive nuclei at time t and λ is the probability of decay per nucleus per unit time.

In contrast, a partial differential equation (PDE) contains partial derivatives. Thus, a PDE is a differential equation in which the dependent variable depends on two or more independent variables [49]. Fick's second law of diffusion is an example of a PDE:

$$\frac{\partial C}{\partial t} = D \frac{\partial^2 C}{\partial x^2}, \tag{8.2}$$

where C is the concentration of the chemical element under investigation and D is a positive constant called the "diffusion coefficient."

To solve a differential equation, we need to find an expression for the dependent variable [i.e., $N(t)$ in Eq. (8.1)] in terms of the independent variable(s) [i.e., t in

© The Author(s), under exclusive license to Springer Nature Switzerland AG 2021   117
M. Petrelli, *Introduction to Python in Earth Science Data Analysis*,
Springer Textbooks in Earth Sciences, Geography and Environment,
https://doi.org/10.1007/978-3-030-78055-5_8

Eq. (8.1)] that satisfies the differential equation. By definition, a differential equation contains derivatives, so finding a solution requires an integration. The general solution of an ODE is the solution that satisfies the differential equation for all initial conditions. It is a combination of functions and one or more constants. A particular solution to an ODE is the solution obtained from the general solution by assigning specific values to the arbitrary constants. ODEs and PDEs can be solved by using both analytical and numerical methods. Although this chapter focuses on using Python to find numerical solutions to differential equations, you should keep in mind that analytical solutions should be explored whenever possible [42].

8.2 Ordinary Differential Equations

As stated above, ODEs contain ordinary derivatives only. The order of an ODE is the order of the highest derivative that appears in the equation. The explicit form of an nth-order ODE can be written as [39]

$$\frac{d^n y}{dx^n} = y^{(n)} = f(x, y, y', y'', \ldots, y^{(n-1)}), \tag{8.3}$$

where f is a known function.

An ODE is linear if the unknown function appears linearly in the equation, otherwise it is nonlinear [39, 42, 49, 60].

Direction fields of first-order ODEs

Direction fields provide an overview of first-order ODE solutions without actually solving the equation. Recall that first-order ODEs are those that can be written in the form [39]

$$y' = f(x, y). \tag{8.4}$$

A direction field is a set of short line segments passing through various, typically grid shaped, points having a slope that satisfies the investigated differential equation at the given point.

To my knowledge, Python offers no straightforward method to plot a simple direction field. However, we can easily develop a function for the scope (code Listing 8.1). Figure 8.1 shows the result of applying code Listing 8.1 to the ODE

$$y' = \frac{x^2}{1 - x^2 - y^2}. \tag{8.5}$$

```
1  import numpy as np
2  from matplotlib import pyplot as plt
3
4  # Direction Field
5  def direction_field(x_min, x_max, y_min, y_max, n_step, lenght, fun
       , ax):
6
7      # this is to avoid RuntimeWarning: divide by zero
8      np.seterr(divide='ignore', invalid='ignore')
9
10     x = np.linspace(x_min, x_max, n_step)
11     y = np.linspace(y_min, y_max, n_step)
12     X, Y = np.meshgrid(x, y)
13     slope = fun(X,Y)
14     slope = np.where(slope == np.inf, 10**3, slope)
15     slope = np.where(slope == -np.inf, -10**3, slope)
16     delta = lenght * np.cos(np.arctan(slope))
17     X1 = X - delta
18     X2 = X + delta
19     Y1 = slope*(X1-X)+Y
20     Y2 = slope*(X2-X)+Y
21     ax.plot([X1.ravel(), X2.ravel()], [Y1.ravel(), Y2.ravel()], 'k-
          ', linewidth=1)
22
23  # Differential equations
24  def my_ode(x, y):
25      dy_dx = x**2 / (1 - x**2 - y**2)
26      return dy_dx
27
28  # Make the plot
29  fig, ax1 = plt.subplots()
30  direction_field(x_min=-2, x_max=2, y_min=-2, y_max=2, n_step=30,
          lenght=0.05, fun=my_ode, ax=ax1)
31  ax1.set_xlabel('x')
32  ax1.set_ylabel('y')
33  ax1.axis('square')
34  ax1.set_title(r"$ {y}' = - \frac{x^2}{1 - x^2 - y^2} $")
```

Listing 8.1 Defining a function to create a direction field

At this point, you should already know the meaning of most of the instructions in code Listing 8.1. Exceptions are the statements at lines 8 and 12. The command at line 8 simply avoids displaying a warning for dividing by zero during the evaluation of the function at line 13. When you divide by zero, the value returned could be inf, $-$ inf, or NaN (i.e., not a number). In the first two cases, the slope is "adjusted" to a "large" number at lines 14 and 15 (i.e., 1000 and -1000, respectively) to plot a vertical segment in the direction field. In the case of NAN, nothing is plotted at the corresponding node of the grid.

At line 12, the command *np.meshgrid(x,y)* returns two coordinate matrices from two coordinate vectors. More specifically, it creates two rectangular arrays: one of x values and one of y values. Combining the resulting matrices, we obtain a rectangular grid of coordinates. This approach is particularly useful when dealing with spatial data.

The *quiver()* method of matplotlib provides an alternative method to investigate the behavior of first-order ODEs. Specifically, it uses the formulation

Fig. 8.1 Result of code
Listing 8.1

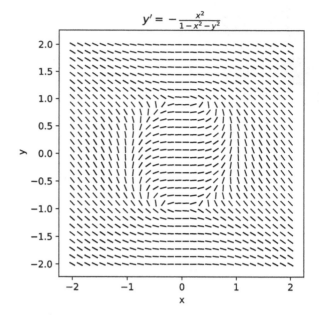

$$y' = -\frac{x^2}{1-x^2-y^2}$$

$$\frac{dx}{dt} = B(x, y),$$
$$\frac{dy}{dt} = A(x, y). \tag{8.6}$$

The advantage of Eq. (8.6) lies in the fact that the *quiver()* function can be used directly to display velocity vectors in $[x, y]$ space. Similarly, the *streamplot()* function visualizes ODE solutions as streamlines. As an example, the code Listing 8.2 implements the direction field and streamlines of the velocity field of

$$\frac{dx}{dt} = x + 2y, \quad \frac{dy}{dt} = -2x. \tag{8.7}$$

Figure 8.2 shows the results of the *quiver()* and *streamplot()* functions in the left and right panels, respectively.

The *quiver()* and *streamplot()* functions can be also used to investigate first-order ODEs in the canonical form $y' = f(x, y)$. This is because $A(x, y)$ and $B(x, y)$ simply derive from the transformation

$$\frac{dy}{dx} = \frac{A(x, y)}{B(x, y)}. \tag{8.8}$$

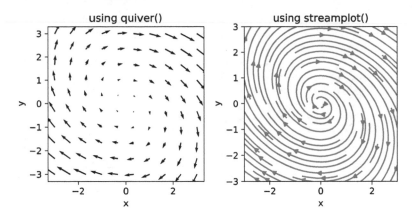

Fig. 8.2 Result of code Listing 8.2

```
1   import numpy as np
2   import matplotlib.pyplot as plt
3
4   x = np.linspace(-3, 3, 10)
5   y = x
6   X, Y = np.meshgrid(x, y)
7
8   dx_dt = X + 2*Y
9   dy_dt = - 2*X
10
11  fig = plt.figure()
12  ax1 = fig.add_subplot(1, 2, 1)
13  ax1.quiver(X, Y, dx_dt, dy_dt)
14  ax1.set_title('using quiver()')
15  ax1.set_xlabel('x')
16  ax1.set_ylabel('y')
17  ax1.axis('square')
18
19  ax2 = fig.add_subplot(1, 2, 2)
20  ax2.streamplot(X, Y, dx_dt, dy_dt)
21  ax2.set_title('using streamplot()')
22  ax2.set_xlabel('x')
23  ax2.set_ylabel('y')
24  ax2.axis('square')
25
26  fig.tight_layout()
```

Listing 8.2 Using the *quiver()* and *streamplot()* methods with first-order ODEs

Consequently, if we assume that $A(x, y) = f(x, y)$ and $B(x, y) = 1$, then Eq. (8.8) reduces to the form $y' = f(x, y)$. As an example, code Listing 8.3 (Fig. 8.3) shows the application of the *quiver()* and streamplot() functions to

$$\frac{dy}{dx} = -y - 2x^2. \tag{8.9}$$

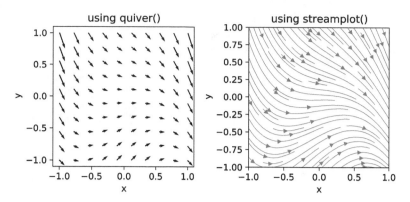

Fig. 8.3 Result of code Listing 8.3

```
1   import numpy as np
2   import matplotlib.pyplot as plt
3
4   x = np.linspace(-1, 1, 10)
5   y = x
6
7   X, Y = np.meshgrid(x, y)
8
9   dx_dt = np.ones_like(X)
10  dy_dt = - Y - 2 * X**2
11
12  # Making plot
13  fig = plt.figure()
14  ax1 = fig.add_subplot(1, 2, 1)
15  ax1.quiver(X, Y, dx_dt, dy_dt)
16  ax1.set_title('using quiver()')
17  ax1.set_xlabel('x')
18  ax1.set_ylabel('y')
19  ax1.axis('square')
20
21  ax2 = fig.add_subplot(1, 2, 2)
22  ax2.streamplot(X, Y, dx_dt, dy_dt, linewidth=0.5)
23  ax2.set_title('using streamplot()')
24  ax2.set_xlabel('x')
25  ax2.set_ylabel('y')
26  ax2.axis('square')
27
28  fig.tight_layout()
```

Listing 8.3 Using the *quiver()* and *streamplot()* functions

8.3 Numerical Solutions of First-Order Ordinary Differential Equations

Equation (8.1) for radioactive decay has an analytical solution of the form

$$N(t) = N_0 e^{-\lambda t} = N_0 e^{-\frac{t}{\tau}} \tag{8.10}$$

where $N(t)$, N_0, λ, and τ are the quantity N at time t, the quantity N at time $t = 0$, the exponential decay constant, and the mean lifetime, respectively. Radioactive decay represents a suitable example to illustrate some of the numerical techniques used to solve both ordinary and partial differential equations.

Euler's method

Euler's method consists of a finite-difference approximation to numerically solve differential equations by taking small finite steps Δt in the parameter t and approximating the function $N(t)$ with the first two terms of its Taylor expansion:

$$\frac{dN}{dt} \approx \frac{N(t + \Delta t) - N(t)}{\Delta t}. \tag{8.11}$$

This approach gives

$$N(t + \Delta t) \approx \frac{dN}{dt} \Delta t + N(t) = -\lambda N(t) \Delta t + N(t) = N(t)(1 - \lambda \Delta t). \tag{8.12}$$

Assuming a decay constant $\lambda = 1.54 \times 10^{-1}$ per billion years (i.e., 1.54×10^{-10} per year) as in the case of the uranium series from ^{238}U to ^{206}Pb, we can code a Python script to solve Eq. (8.1) (code Listing 8.4) and compare the analytical solution with the numerical solution (Fig. 8.4). The deviation from the expected value of Euler's method (i.e., the error) is a function of Δt.

In addition, Euler's method is affected by an intrinsic issue, potentially leading to a progressive amplification of the error because it evaluates the derivatives only at the beginning of the investigated interval (i.e., Δt). If the derivative at the beginning of the step is systematically incorrect, either too high or too low, then the numerical solution will suffer from the same systematic error (Fig. 8.4).

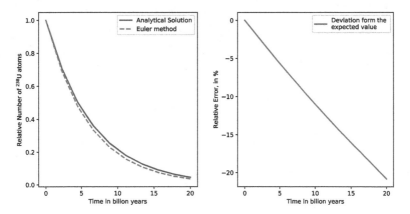

Fig. 8.4 Result of code Listing 8.4

Note that, in this specific case, we use a quite large Δt value to highlight the limits of Euler's method. Reducing Δt significantly improves the accuracy. However, for the general case, improving the accuracy requires estimating the derivative at more than one point in the investigated interval.

```python
1   import matplotlib.pyplot as plt
2   import numpy as np
3
4   # Euler Method
5   def euler_method(n0, decay_const, t_final, n_t_steps):
6       iterations = n_t_steps
7       delta_t = t_final/n_t_steps
8       t1 = np.linspace(0, iterations*delta_t, iterations)
9       n1 = np.zeros(t1.shape, float)
10      n1[0]=n0
11      for i in range(0,len(t1)-1):
12          n1[i+1] = n1[i] * (1 - decay_const * delta_t )
13      n1r = n1/n0
14      return n1, n1r, t1
15
16  ne, ner, te = euler_method(n0=10000, decay_const=1.54e-1,
        t_final=20, n_t_steps=10)
17
18  #Analitical solution ...in the same points of the Euler method
19  def analytical_solution(n0, decay_const, t_final, n_t_steps):
20
21      intermediate_points = n_t_steps
22      delta_t = t_final/n_t_steps
23      t2 = np.linspace(0, intermediate_points*delta_t,
        intermediate_points)
24      n2 = n0 * np.exp(-decay_const * t2 )
25      n2r = n2/n0
26      return n2, n2r, t2
27
28  na, nar, ta = analytical_solution(n0=10000, decay_const=1.54e
        -1, t_final=20, n_t_steps=10)
29
30  euler_rel_error = 100*(ne-na)/na
31
32  fig = plt.figure()
33  ax1 = fig.add_subplot(1, 2, 1)
34  ax1.plot(te, ner, linestyle="-", linewidth=2, label='Euler
        method')
35  ax1.plot(ta, nar, linestyle="--", linewidth=2, label='
        Analytical Solution')
36  ax1.set_ylabel('Relative Number of $^{238}$U atoms')
37  ax1.set_xlabel('time in bilion years')
38  ax1.legend()
39
40  ax2 = fig.add_subplot(1, 2, 2)
41  ax2.plot(te, euler_rel_error, linestyle="-", linewidth=2, label
        ='Deviation formthe \nexpected value')
42  ax2.set_ylabel('Relative Error, in %')
43  ax2.set_xlabel('time in bilion years')
44  ax2.legend()
```

Listing 8.4 Implementing Euler's method in Python

The scipy.integrate.ode class

The scipy.integrate.ode class is generic interface class for numeric solutions of ODEs. Available integrators are

1. a real-valued variable-coefficient ODE solver (i.e., *vode*);
2. a complex-valued variable-coefficient ODE solver with fixed-leading-coefficient
   implementation (i.e., *zvode*);
3. a real-valued variable-coefficient ODE solver with fixed-leading-coefficient
   implementation (i.e., *lsoda*);
4. an explicit Runge–Kutta method of order (4)5 (i.e., *dopri5*);
5. an explicit Runge–Kutta method of order 8(5, 3) (i.e., *dopri853*).

Please refer to more specialized books for a detailed description of these methods [41, 48, 50].

```
1   import matplotlib.pyplot as plt
2   import numpy as np
3   from scipy.integrate import ode
4
5   # using scipy.integrate.ode
6   def ode_sol(n0, decay_const, t_final, n_t_steps):
7       intermediate_points = n_t_steps
8       t3 = np.linspace(0,t_final, intermediate_points)
9       n3 = np.zeros(t3.shape, float)
10      def f(t, y, decay_const):
11          return  - decay_const * y
12      solver = ode(f).set_integrator('dopri5') # runge-kutta of order
          (4)5
13      y0 = n0
14      t0 = 0
15      solver.set_initial_value(y0, t0)
16      solver.set_f_params(decay_const)
17      k=1
18      n3[0] = n0
19      while solver.successful() and solver.t < t_final:
20          n3[k] = solver.integrate(t3[k])[0]
21          k += 1  # k = k + 1
22      n3r = n3 / n0
23      return n3, n3r, t3
24
25  # Analytical solution
26  na, nar, ta = analytical_solution(n0=10000, decay_const=1.54e-1,
          t_final=20, n_t_steps=10)
27  # Euler method
28  ne, ner, te = euler_method(n0=10000, decay_const=1.54e-1, t_final
          =20, n_t_steps=10)
29  nuler_rel_error = 100*(ne-na)/na
30  # runge-kutta of order (4)5
31  n_ode, n_oder, tode = ode_sol(n0=10000, decay_const=1.54e-1,
          t_final=20, n_t_steps=10)
32  ode_rel_error = 100*(n_ode - na) / na
33
34  # Make the plot
35  fig = plt.figure(figsize=(8,5))
36  ax1 = fig.add_subplot(1, 2, 1)
37  ax1.plot(ta, nar, linestyle="-", linewidth=2, label='Analytical
          Solution', c='#ff464a')
38  ax1.plot(te, ner, linestyle="--", linewidth=2, label='Euler method'
          , c='#4881e9')
39  ax1.plot(tode, n_oder, linestyle="--", linewidth=2, label='Runge-
          Kutta of order (4)5', c='#342a77')
40  ax1.set_ylabel('Relative Number of $^{238}$U atoms')
41  ax1.set_xlabel('Time in bilion years')
```

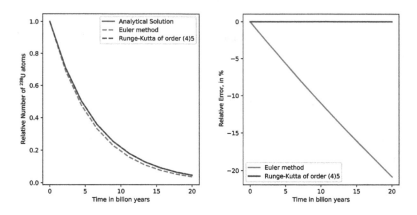

Fig. 8.5   Result of code Listing 8.5

```
42  ax1.legend()
43
44  ax2 = fig.add_subplot(1, 2, 2)
45  ax2.plot(te, euler_rel_error, linestyle="-", linewidth=2, c='#4881
        e9', label='Euler method')
46  ax2.plot(tode, ode_rel_error, linestyle="-", linewidth=2, c='#342
        a77', label='Runge-Kutta of order (4)5')
47  ax2.set_ylabel('Relative Error, in %')
48  ax2.set_xlabel('Time in bilion years')
49  ax2.legend()
50
51  fig.tight_layout()
```
Listing 8.5   Euler's method versus Runge–Kutta of order (4)5

I now show how to apply the scipy.integrate.ode class to the real case of radioactive decay. The code Listing 8.5 (Fig. 8.5) concerns the application of the explicit Runge–Kutta method of order (4)5 to the equations investigated and compares the results with those of Euler's method.

8.4   Fick's Law of Diffusion—A Widely Used Partial Differential Equation

As originally reported by [47], the rate of transfer of a substance diffusing through a unit area is proportional to the concentration gradient measured normal to the area:

$$F = -D\frac{\partial C}{\partial x}, \tag{8.13}$$

where F is the rate of transfer per unit area, C the concentration of the diffusing substance, x the spatial coordinate measured normal to the area, and D is the diffusion coefficient. In some cases, such as diffusion in dilute solutions, D can reasonably be considered as a constant, whereas, in other cases, such as diffusion in high polymers, it depends strongly on concentration [45]. Equation (8.13) is universally known as the first Fick law or the first law of diffusion [45]. The units of C, F, x, and D are concentration (e.g., mol $\cdot$ m^{-3}), concentration per unit time (e.g., mol $\cdot$ m$^{-3}\cdot$ s^{-1}, length (e.g., m), and length squared divided by time (e.g., m$^2 \cdot$ s^{-1}), respectively [55]. In the case of one-dimensional processes characterized by a constant D, Eq. (8.13) can be written as the one-dimensional second Fick's law or second law of diffusion [45]:

$$\frac{\partial C}{\partial t} = D\frac{\partial^2 C}{\partial x^2}. \tag{8.14}$$

For constant D and specific geometries, Eq. (8.2) can be solved analytically [45]. In all other cases the problem requires a numerical solution.

Analytical solutions

If the diffusion coefficient is constant, analytical solutions to the diffusion equation can be obtained for a variety of initial conditions and boundary conditions [45]. As an example, the solution for a diffusing substance initially confined in a finite region,

$$C = C_0, \quad x < 0, \quad C = 0, \quad x \geq 0, \quad t = 0, \tag{8.15}$$

can be written written in the form [45]

$$C(x, t) = \frac{1}{2}C_0 \text{erfc}\left(\frac{x}{2\sqrt{Dt}}\right), \tag{8.16}$$

where erfc() is the complementary error function defined as $1 - \text{erf}()$:

$$\text{erfc}(x) = 1 - \text{erf}(x) = \frac{2}{\sqrt{\pi}} \int_x^\infty e^{-t^2} dt. \tag{8.17}$$

More details about the error function are given in Sect. 9.2.
Code Listing 8.6 (Fig. 8.6) implements Eq. (8.16) in Python.

```
1   import numpy as np
2   from scipy import special
3   import matplotlib.pyplot as plt
4
5   def plane_diff_1d(t, D, x0=0, xmin=-1, xmax=1, c_left=1, c_right=0,
        num_points=200):
6
7       n = num_points
8       x = np.linspace(xmin, xmax, n)
9       delta_c = c_left - c_right
10
11      c0 = np.piecewise(x, [x < x0, x >= x0], [c_left, c_right])
```

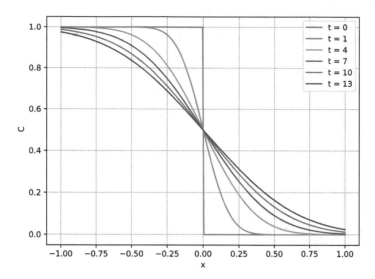

Fig. 8.6 Result of code Listing 8.6

```
12        c = 0.5 * delta_c * (special.erfc((x - x0)/(2 * np.sqrt(D * t)
          )))
13
14        return x, c, c0
15
16  D = 0.01 # Diffusion coefficient
17
18  fig, ax = plt.subplots()
19
20  for t in range(1, 14, 3):
21
22      x, c, c0 = plane_diff_1d(t=t, D=D)
23      if t==1:
24          leg = "t = " + str(t)
25          plt.plot(x, c0, label="t = 0")
26      leg = "t = " + str(t)
27      ax.plot(x, c, label=leg)
28
29  ax.grid()
30  ax.set_xlabel('x')
31  ax.set_ylabel('C')
32  ax.legend()
```

Listing 8.6 Analytical solution of plane diffusion

Numerical solution for constant D

The simplest way to discretize Eq. (8.2) is by using finite differences [51]:

$$\frac{C_j^{n+1} - C_j^n}{\Delta t} = D \left[\frac{C_{j+1}^n - 2C_j^n + C_{j-1}^n}{(\Delta x)^2} \right], \tag{8.18}$$

where n and j represent the time and space domain, respectively. This scheme, called "Forward-Time Central-Space" (FTCS), uses Euler's method and a central-difference scheme to approximate the derivatives in time and space, respectively. For more details about the theory behind numerical schemes to solve PDEs, please refer to more specialized books [51, 52, 54].

In Python, Eq. (8.18) can be easily implemented by using code Listing 8.7. The finite-difference scheme given by Eq. (8.18) is stable under the condition

$$\frac{2D\Delta t}{(\Delta x^2)} \leq 1. \tag{8.19}$$

Figure 8.7 compares the analytical solution with the numerical solution.

The implementation of the FTCS scheme (i.e., lines 1–16 of the code Listing 8.7) could be improved by more fully exploiting numpy [51]. In particular, the loop at lines 4–6 could be replaced by a single line of code using vectorial notation [line 4 of code Listing 8.8 [51]].

```
1   def ftcs(u, D, h, dt):
2
3       d2u_dx2 = np.zeros(u.shape, float)
4       for i in range(1,len(u)-1):
5           # Central difference scheme in space
6           d2u_dx2[i] = (u[i+1] - 2*u[i] + u[i-1]) / h**2
7
8       # Neuman boundary conditions at i=0 and i=len(u)-1
9       i=0
10      d2u_dx2[i] = (u[i+1] - 2 * u[i] + u[i]) / h**2
11      i=len(u)-1
12      d2u_dx2[i] = (u[i] - 2 * u[i] + u[i-1]) / h**2
13
14      # Euler method for the time domain
15      u1 = u + dt * D * d2u_dx2
16      return u1
17
18  dt = 0.001   #step size of time
19  tf = 3
20
21  def compute_d_const(u, d, h, dt, tf):
22
23      nsteps = tf/dt
24      u1 = u
25      for i in range(int(nsteps)):
26          u1 =  ftcs(u1, D, h, dt)
27      return u1
28
29  x, c, c0 = plane_diff_1d(t=tf, D=D)
30
31  h = x[1] - x[0] #step size of the 1D space
32  u = c0 # intial conditions
33  c1 = compute_d_const(u, D, h, dt, tf)
34
35  fig, ax = plt.subplots()
36  ax.plot(x,c0, label='initial conditions')
37  ax.plot(x,c,'y', label='analytical solution')
38  ax.plot(x,c1,'r--', label='numerical solution')
39  ax.set_xlabel('x')
40  ax.set_ylabel('C')
41  ax.legend()
```

Listing 8.7 Plane diffusion by finite-difference method

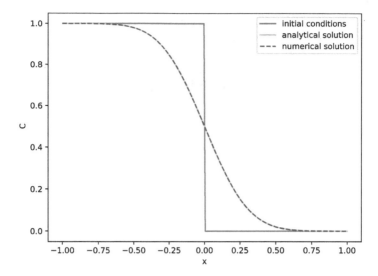

Fig. 8.7   Result of code Listing 8.7

```
 1  def numpy_ftcs(u, D, h, dt):
 2
 3      d2u_dx2 = np.zeros(u.shape, float)
 4      d2u_dx2[1:-1] = (u[2:] - 2 * u[1:-1] + u[:-2]) / h**2
 5
 6      # Neuman boundary conditions at i=0 and i=len(u)-1
 7      i = 0
 8      d2u_dx2[i] = (u[i+1] - 2 * u[i] + u[i]) / h**2
 9      i = len(u)-1
10      d2u_dx2[i] = (u[i] - 2 * u[i] + u[i-1]) / h**2
11
12      # Euler method for the time domain
13      u1 = u + dt * D * d2u_dx2
14      return u1
```

Listing 8.8   Using the vectorial notation for the FTCS scheme

In petrology and volcanology, the process of chemical diffusion is often used to constrain the residence times of crystals in a volcanic plumbing system before an eruption [44]. For example, [43] report a formulation to model diffusion of Mg in plagioclase, while also accounting for the influence of anorthite absolute values and gradients on chemical potentials and diffusion coefficients [44]. The rationale behind the formulation provided by [44] comes from diffusive fluxes of trace elements being strongly coupled to major element concentration gradients (i.e., multi-component diffusion).

In the following, I provide a Python implementation for the problem reported by [44]. The code consists of

1. analytically determining Mg and An contents on zoned plagioclases;

2. constraining the boundary conditions (e.g., equilibrium at the rims);
3. defining the initial and equilibrium profiles;
4. estimating the diffusion coefficient for Mg;
5. solving the time-dependent form of the diffusion equation by using finite differences.

The analytical determinations for An are typically done by electron-probe microanalysis (EPMA). Magnesium can be determined either by EPMA or laser ablation inductively coupled plasma mass spectrometry (LA-ICP-MS).

For example, Fig. 8.8 shows a rim-to-rim MgO profile, analyzed by EPMA, on the plagioclase labelled 4202-1 Pl1 by [53].

Following the approach proposed by [43], the dependence of Mg trace element partitioning between plagioclase and melt on the anorthite content is approximated by [43]

$$RT \ln \frac{C_{Mg}^{Pl}}{C_{Mg}^{l}} = A X_{An} + B, \qquad (8.20)$$

where X_{An}, C_{Mg}^{Pl}, C_{Mg}^{l}, T, and R are the anorthite molar fraction, the concentration of Mg in plagioclase, the concentration of Mg in the liquid, the temperature, and the universal gas constant, respectively [43]. For the A and B parameters, [53] proposed $A = -21882$ and $B = -26352$.

To model the diffusive process, the initial and equilibrium profiles have been estimated starting from Eq. (8.20). Specifically, the initial profile is defined by the melt concentration in equilibrium with the crystal core [53]. Both the initial and equilibrium profiles are calculated by using Eq. (8.20) (i.e., MgO is 7.8 and 8.4 wt%, respectively).

As boundary conditions, the crystal rims in contact with the surrounding melt are open. This means that Mg values at the rims are those of the equilibrium profile, where I used the formulation for the diffusion coefficient reported by [43]:

$$D_{Mg} = \left[2.92 \times 10^{-4.1 X_{An} - 3.1} \exp \left(\frac{-266\,000}{RT} \right) \right] \times 10^{12}. \qquad (8.21)$$

The time-dependent form of the diffusion equation developed by [43] for Mg in a plagioclase is

$$\frac{\partial C_{Mg}}{\partial t} = \left(D_{Mg} \frac{\partial^2 C_{Mg}}{\partial x^2} + \frac{\partial C_{Mg}}{\partial x} \frac{\partial D_{Mg}}{\partial x} \right) \qquad (8.22)$$
$$- \frac{A}{RT} \left(D_{Mg} \frac{\partial C_{Mg}}{\partial x} \frac{\partial X_{An}}{\partial x} + C_{Mg} \frac{\partial D_{Mg}}{\partial x} \frac{\partial X_{An}}{\partial x} + D_{Mg} C_{Mg} \frac{\partial^2 X_{An}}{\partial x^2} \right).$$

Code Listing 8.9 shows the finite-difference approximation of Eq. (8.22). In accordance with [53], I used in code Listing 8.9 the FTCS scheme given in Eq. (8.18). For first-order derivatives, the explicit central scheme in space is

$$\frac{\partial C}{\partial x} \approx \frac{C_{j+1}^n - C_{j-1}^n}{2\Delta x}. \tag{8.23}$$

The notation is the same as that used in Eq. (8.7).

```
1   import numpy as np
2   import matplotlib.pyplot as plt
3   import pandas as pd
4
5   # Model parameters
6   T = 1200.0 # Temperature in Celsius
7   dx = 4.12 # average distance in micron among the analyses
8   dt = 0.9 * 1e4
9   RT = 8.3144 * (T + 273.15)
10  R = dt / dx ** 2
11
12  # Initial Conditions
13  my_dataset = pd.read_excel('Moore_Phd.xlsx')
14  my_distance = my_dataset.Distance.values
15  Mg_C = my_dataset.MgO.values
16  An = my_dataset.An_mol_percent.values
17  An = An / 100
18  An_unsmoothed = An
19  An_smoothed = np.full(len(An),0.)
20
21  # Smooting the An profile to avoid numerical artifacts
22  D_smoot = np.full(len(An),0.0005)
23  for i in range(2):
24      An_smoothed[1:-1] =  An_unsmoothed[1:-1] + R * D_smoot[1:-1] *
            (An_smoothed[2:] - 2 * An_unsmoothed[1:-1] + An_unsmoothed
            [:-2])
25      An_smoothed[0] = An[0]
26      An_smoothed[len(An)-1] = An[len(An)-1]
27      an_unsmoothed = An_smoothed
28
29  D_Mg = 2.92 * 10**(-4.1 * An_smoothed - 3.1)*np.exp(-266 * 1e3/RT)
            *1e12  # Eq. 8 in Costa et al., 2003
30
31  fig, ax = plt.subplots(figsize=(7,5))
32
33  # Initial and Equilibrium Profiles
34  A = - 21882
35  B = - 26352
36  K = np.exp((A*An_smoothed+B)/RT) # Eq. 8 in Moore et al., 2014
37  c_eq = 8.4 * K
38  c_init = 7.8 * K
39  ax.plot(my_distance, c_eq, linewidth=2, color='#ff464a', label ='
            Equilibrium Profile')
40  ax.plot(my_distance, c_init,linewidth=2,  color='#342a77', label='
            Initial Profile')
41
42  # The numerical solution start here
43  colors = ['#4881e9','#e99648','#e9486e']
44  t_final_weeks = np.array([4,10,21])
45
46  for t_w, color in zip(t_final_weeks,colors):
47
48      C_Mg_new = np.full(len(c_eq),0.)
49      d2An = np.full(len(c_eq),0.)
50      d2C_Mg = np.full(len(c_eq),0.)
```

```
51      dD_Mg = np.full(len(c_eq),0.)
52      dC_Mg = np.full(len(c_eq),0.)
53      dAn = np.full(len(c_eq),0.)
54
55      C_Mg = c_init
56      t_final = int(604800 * t_w/dt)
57      for i in range(t_final):
58          # boundary conditions: Rims are at equilibrium with melt
59          C_Mg_new[0] = c_eq[0]
60          C_Mg_new[len(c_eq)-1] = c_eq[len(c_eq)-1]
61
62          # Finite difference sol. of Eq. 7 in Costa et al., 2003
63          d2An[1:-1] = (An_smoothed[2:] - 2 * An_smoothed[1:-1] +
        An_smoothed[:-2])
64          d2C_Mg[1:-1] = C_Mg[2:] - 2 * C_Mg[1:-1] + C_Mg[:-2]
65          dD_Mg[1:-1] =  (D_Mg[2:]-D_Mg[:-2])/2
66          dC_Mg[1:-1] = (C_Mg[2:]-C_Mg[:-2])/2
67          dAn[1:-1] =  (An_smoothed[2:]-An_smoothed[:-2])/2
68
69          C_Mg_new[1:-1] = C_Mg[1:-1] + R * ( (D_Mg[1:-1] * d2C_Mg
        [1:-1] + dD_Mg[1:-1] * dC_Mg[1:-1]) - (A/RT) * (D_Mg[1:-1] *
        dC_Mg[1:-1] * dAn[1:-1]  +  C_Mg[1:-1] * dD_Mg[1:-1] * dAn
        [1:-1] + D_Mg[1:-1] * C_Mg[1:-1] * d2An[1:-1]) )
70          C_Mg = C_Mg_new
71      ax.plot(my_distance, C_Mg_new, linestyle='--', linewidth=1,
        label= str(t_w) + ' weeks at 1200 Celsius deg.')
72
73  ax.scatter(my_distance, Mg_C, marker='o', c='#c7ddf4', edgecolors=
        'k', s=50, label='Analytical Deteminations', zorder=100, alpha
        =0.7)
74  ax.set_ylim(0.19,0.27)
75
76  time_sec = t_final * dt
77  time_weeks = time_sec / 604800
78  ax.legend(title=r'$\bf{4202\_1-P11}$ (Moore et al., 2014)', ncol=2,
        loc='lower center')
79  ax.set_xlabel(r'Distance [$\mu m$]')
80  ax.set_ylabel('MgO  [wt %]')
81  fig.tight_layout()
```

Listing 8.9 Implementation of Eq. (8.22) in Python. Data are from [53]

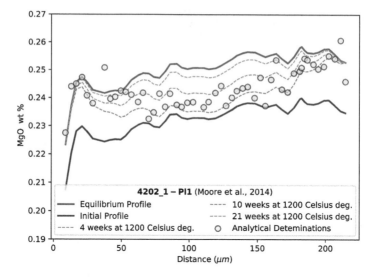

Fig. 8.8   Result of code Listing 8.9

Part IV
Probability Density Functions and Error Analysis

Chapter 9
Probability Density Functions and Their Use in Geology

9.1 Probability Distribution and Density Functions

Everitt [65] defines the probability distribution for discrete random variables as the mathematical formula that gives the probability for the variables to take on any given value. For a continuous random variable, this function returns the likelihood of an outcome and is graphically described by a curve in the (x, y) plane. For a specific interval $[x_1, x_2]$, the area under the curve (i.e., the definite integral) provides the probability that the investigated variable falls within $[x_1, x_2]$ [65]. The term "probability density" also refers to a probability distributions [65].

Definition: A probability density function (PDF) is a function associated with a continuous random variable whose value at any point in the sample space (i.e., the set of values possible for the random variable) is an estimate of the likelihood of occurrence for that specific value. All PDFs share the following properties and indexes [68]:

- the PDF is normalized when $\int_{-\infty}^{\infty} \mathrm{PDF}(x)dx = 1$;
- the probability that x lies between the values $x_1 \leqslant x_2$ is $P(x) = \int_{x_1}^{x_2} \mathrm{PDF}(x)dx$;
- the mean μ is $\int_{-\infty}^{\infty} x\mathrm{PDF}(x)dx$;
- the median M_e is given by $\int_{-\infty}^{M_e} \mathrm{PDF}(x)dx = \frac{1}{2}$;
- the variance σ^2 is $\int_{-\infty}^{\infty} (x - \mu)^2 \mathrm{PDF}(x)dx$;
- the skewness μ_3 is $\int_{-\infty}^{\infty} (x - \mu)^3 \mathrm{PDF}(x)dx$.

The second point tells us that, solving a definite integral in the interval $[x_1, x_2]$ for a variable x describing a geological process with a known PDF, we define the probability for the occurrence of the process between x_1 and x_2. A specific example is given later in the chapter. Unfortunately, the PDF is rarely known *a priori*. Under

© The Author(s), under exclusive license to Springer Nature Switzerland AG 2021
M. Petrelli, *Introduction to Python in Earth Science Data Analysis*,
Springer Textbooks in Earth Sciences, Geography and Environment,
https://doi.org/10.1007/978-3-030-78055-5_9

specific conditions, our measurements could follow a known PDF. For example, the different formulations of the Central Limit Theorem tell us the circumstances under which the estimates of a sample mean converge to a normal distribution (Sect. 9.6).

9.2 The Normal Distribution

Normal probability density function

The normal distribution is a bell-shaped PDF that occurs naturally in many situations. For example, it is often applied to calibrate analytical devices, in error propagation (Sect. 10), and, generally speaking, to interprete data sets resulting from a field campaign (e.g., as a consequence of the Central Limit Theorem; see Sect. 9.6). The normal probability density function (PDF$_N$) is defined as follows:

$$PDF_N(x, \mu, \sigma) = \frac{1}{\sqrt{2\pi\sigma^2}} e^{-\frac{(x-\mu)^2}{2\sigma^2}}, \tag{9.1}$$

where μ and σ are the mean and the standard deviation, respectively. The following list gives the main characteristics of the normal distribution:

- A normal distribution is bell-shaped with points of inflection at $\mu \pm \sigma$.
- The mean, mode, and median are all the same.
- The curve is symmetric about its the center (i.e., around the mean μ).
- All normal curves are non-negative for all x.
- Exactly half of the values are to the left of the center and exactly half of the values are to the right.
- The limit of PDF$_N(x, \mu, \sigma)$ as x goes to positive or negative infinity is zero.
- The height of any normal curve is maximal at $x = \mu$.
- The total area under the curve is unity.
- The shape of any normal curve depends on its mean μ and standard deviation σ (see code Listing 9.1 and Fig. 9.1).
- The standardized normal PDF has a standard deviation of unity and a mean of zero.

The ScyPy library provides the PDF for a normal, or Gaussian, distribution through the function *scipy.stats.norm.pdf()*, but we can also define it by using the *def* statement (i.e., define our own function; see code Listing 9.1 and Fig. 9.1).

```
1   from scipy.stats import norm
2   import matplotlib.pyplot as plt
3   import numpy as np
4
5
6   # I'm going to define my normal PDF...
7   def normal_pdf(x, mean, std):
8       return 1/(np.sqrt(2*np.pi*std**2))*np.exp(-0.5*((x - mean)
            **2)/(std**2))
9
10
11  x = np.arange(-12, 12, .001)
12
13  pdf1 = normal_pdf(x, mean=0, std=2)
14
15  #the built-in norm PDF in scipy.stats
16  pdf2 = norm.pdf(x, loc=0, scale=2)
17
18  fig = plt.figure(figsize=(7,9))
19  ax1 = fig.add_subplot(3, 1, 1)
20  ax1.plot(x,pdf1, color='#84b4e8', linestyle="-", linewidth=6,
            label="My normal PDF")
21  ax1.plot(x,pdf2, color='#ff464a', linestyle="--", label="norm.
            pdf() in scipy.stats ")
22  ax1.set_xlabel("x")
23  ax1.set_ylabel("PDF(x)")
24  ax1.legend(title = r"Normal PDF with $\mu$=0 and 1$\sigma$=2")
25
26
27  ax2 = fig.add_subplot(3, 1, 2)
28  for i in [1, 2, 3]:
29      y = normal_pdf(x,0,i)
30      ax2.plot(x, y, label=r"$\mu$ = 0, 1$\sigma$ = " + str(i))
31  ax2.set_xlabel("x")
32  ax2.set_ylabel("PDF(x)")
33  ax2.legend()
34
35  ax3 = fig.add_subplot(3, 1, 3)
36  for i in [-3, 0, 3]:
37      y = normal_pdf(x, i, 1)
38      ax3.plot(x, y, label=r"$\mu$ = " + str(i) + ", 1$\sigma$ =
            1")
39  ax3.set_xlabel("x")
40  ax3.set_ylabel("PDF(x)")
41  ax3.legend()
42
43  fig.tight_layout()
```

Listing 9.1 The normal PDF

To get the probability of occurrence for a normally distributed entity x between x_1 and x_2, we must solve the definite integral

$$P(x_1 \leq x \leq x_2) = \int_{x_1}^{x_2} \text{PDF}_\text{N}(x, \mu, \sigma)dx = \frac{1}{\sqrt{2\pi\sigma^2}} \int_{x_1}^{x_2} e^{-\frac{(x-\mu)^2}{2\sigma^2}} dx. \quad (9.2)$$

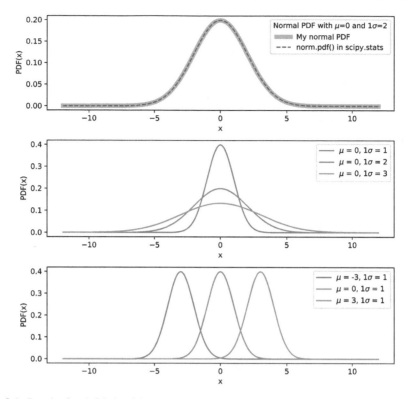

Fig. 9.1 Result of code Listing 9.1

Equation 9.2 has no analytical solution, but given its importance, mathematicians have developed a specific function to solve it: the error function.

The error function $\mathrm{erf}(x)$ is defined as

$$\mathrm{erf}(x) = \frac{2}{\sqrt{\pi}} \int_0^x e^{-t^2} dt. \tag{9.3}$$

Consequently, the solution of the definite integral in Eq. (9.2) (i.e., a normal PDF in the interval $[x_1, x_2]$ with $x_1 \leq x_2$) has the form

$$P(x_1 \leq x \leq x_2) = \frac{1}{\sqrt{2\pi\sigma^2}} \left[\mathrm{erf}\left(\frac{x_2 - \mu}{\sqrt{2\pi\sigma^2}}\right) - \mathrm{erf}\left(\frac{x_1 - \mu}{\sqrt{2\pi\sigma^2}}\right) \right]. \tag{9.4}$$

Also, Eq. (9.2) can be solved numerically by using the techniques given in Chap. 7. The code Listing 9.2 gives the solution of Eq. (9.2) using both Eq. (9.4) and the numerical methods *trapz()* [Eq. (7.14)] and *sims()* [Eq. (7.15)] available in scipy.integrate.

```
1   from scipy.stats import norm
2   import numpy as np
3   from scipy import special
4   from scipy import integrate
5
6   def integrate_normal(x1, x2, mu, sigma):
7       sup = 0.5*((special.erf((x2-mu)/(sigma*np.sqrt(2))))-(
            special.erf((x1-mu)/(sigma*np.sqrt(2)))))
8       return sup
9
10  my_mu = 0
11  my_sigma = 1
12
13  my_x1 = 0
14  my_x2 = my_sigma
15
16  # The expected value is equal to 0.3413...
17  my_sup = integrate_normal(x1= my_x1, x2= my_x2, mu = my_mu,
            sigma = my_sigma)
18
19  x = np.arange(my_x1, my_x2, 0.0001)
20  y =  norm.pdf(x, loc=my_mu, scale= my_sigma) # normal_pdf(x,
            mean = my_mu, std = my_sigma)
21
22  sup_trapz = integrate.trapz(y,x)
23  sup_simps = integrate.simps(y,x)
24
25  print("Solution Using erf: {:.9f}".format(my_sup))
26  print("Using the trapezoidal rule, trapz: {:.10f}".format(
            sup_trapz))
27  print("Using the composite Simpson rule, simps: {:.10f}".format
            (sup_simps))
28
29  '''
30  Output:
31  Solution Using erf: 0.341344746
32  Using the trapezoidal rule, trapz: 0.3413205476
33  Using the composite Simpson rule, simps: 0.3413205478
34  '''
```

Listing 9.2 Solving Eq. (9.2) using Eq. (9.4) and numerical methods

Generating a normal sample distribution

The function *numpy.random.normal(loc=0.0, scale=1.0, size=None)* generates random samples from a normal distribution (code Listing 9.3). Generating a random sample with a specific distribution has many applications. This chapter uses it to demonstrate the properties of the various distributions. Random samples can be also used to perform modeling in Earth Sciences. For example, random sampling is at the foundations of error propagation in the Monte Carlo method (Sect. 10.4).

```
1   import numpy as np
2   from scipy.stats import norm
3   import matplotlib.pyplot as plt
4
5   mu = 0 # mean
```

```
6   sigma = 1 # standard deviation
7   normal_sample = np.random.normal(mu, sigma, 15000)
8
9   # plot the histogram of the sample distribution
10  fig, ax = plt.subplots()
11  ax.hist(normal_sample, bins='auto', density=True, color='#c7ddf4'
        , edgecolor='#000000', label='Random sample with normal
        distribution')
12
13  # probability density function
14  x = np.arange(-5,5, 0.01)
15  normal_pdf = norm.pdf(x, loc= mu, scale = sigma)
16  ax.plot(x, normal_pdf, color='#ff464a', linewidth=1.5, linestyle=
        '--', label=r'Normal PDF with $\mu$=0 and 1$\sigma$=1')
17  ax.legend(title='Normal Distribution')
18  ax.set_xlabel('x')
19  ax.set_ylabel('Probability Density')
20  ax.set_xlim(-5,5)
21  ax.set_ylim(0,0.6)
22
23  # Descriptive statistics
24  aritmetic_mean = normal_sample.mean()
25  standard_deviation = normal_sample.std()
26
27  print('Sample mean equal to {:.4f}'.format(aritmetic_mean))
28  print('Sample standard deviation equal to {:.4f}'.format(
        standard_deviation))
29
30  '''
31  Output: (your results will be sighly different because of the
        pseudo-random nature of the distribution)
32  Sample mean equal to -0.0014
33  Sample standard deviation equal to 1.0014
34  '''
```

Listing 9.3 Generating a random sample with normal distribution ($\mu = 0$ and $1\sigma = 1$) and a normal PDF having the same μ and 1σ as the random sample.

Note that Monte Carlo simulations form the basis of many geological studies involving estimates of uncertainties in the field of mineral exploration mapping [83], slope stability [80], and groundwater hydrology [63].

Code Listing 9.3 shows how to generate a random sample of 15 000 elements characterized by $\mu = 0$ and $\sigma = 1$. Also, code Listing 9.3 produces a normal PDF with the same μ and 1σ as the random sample (Fig. 9.2).

9.3 The Log-Normal Distribution

The log-normal (or lognormal) distribution is a continuous probability distribution of a random variable whose logarithm is normally distributed. The log-normal distribution is often invoked as a fundamental rule in geology [62]. Despite its pitfalls, it remains widely used by geologists today [73]. The PDF for a log-normal distribution is given by

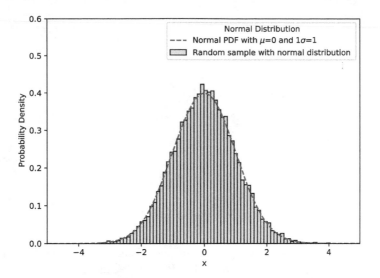

Fig. 9.2 Result of code Listing 9.3

$$\mathrm{logPDF_N}(x, \mu_n, \sigma_n) = \frac{1}{x} \frac{1}{\sqrt{2\pi\sigma_n^2}} \exp\left\{-\frac{[\log(x) - \mu_n]^2}{2\sigma_n^2}\right\}, \qquad (9.5)$$

where μ_n and σ_n are the mean and the standard deviation of the normal distribution and are obtained by calculating the natural logarithm of the random variable.

```
1   import matplotlib.pyplot as plt
2   import numpy as np
3   from scipy.stats import norm, lognorm
4
5   colors = ['#342a77', '#ff464a', '#4881e9']
6   normal_mu = [0,0.5,1]
7   normal_sigma = [0.5,0.4,0.3]
8   x = np.arange(0.001, 7, .001) # for the log-normal PDF
9   x1 = np.arange(-2.5, 2.5, .001)  # for the normal PDF
10
11  fig, (ax1, ax2) = plt.subplots(nrows = 2, ncols = 1, figsize =
       (8,9))
12
13  for mu_n, sigma_n, color in zip(normal_mu, normal_sigma, colors):
14      lognorm_pdf = lognorm.pdf(x, s=sigma_n, scale=np.exp(mu_n))
15      r = lognorm.rvs(s=sigma_n, scale=np.exp(mu_n), size=15000)
16      ax1.plot(x, lognorm_pdf, color=color, label=r"$\mu_n$ = " +
       str(mu_n) + r" - $\sigma_n$ = " + str(sigma_n))
17      ax1.hist(r, bins='auto', density=True, color=color, edgecolor
       ='#000000', alpha=0.5)
18      logr= np.log(r)
19      normal_pdf = norm.pdf(x1, loc= mu_n, scale = sigma_n)
20      ax2.plot(x1, normal_pdf, color=color, label=r"$\mu_n$ = " +
       str(mu_n) + r" - $\sigma_n$ = " + str(sigma_n))
21      ax2.hist(logr, bins='auto', density=True, color=color,
       edgecolor='#000000', alpha=0.5)
22      my_mu = logr.mean()
```

```
23        ax2.axvline(x=my_mu, color=color, linestyle="--", label=r"
          calculated $\mu_n$ = " + str(round(my_mu,3)))
24        my_sigma = logr.std()
25        print("Expected mean: " + str(mu_n) + " - Calculated mean: "
          + str(round(my_mu,3)))
26        print("Expected std.dev.: " + str(sigma_n) + " - Calculated
          std.dev.: " + str(round(my_sigma,3)))
27
28     ax1.legend(title="log-normal distributions")
29     ax1.set_xlabel('x')
30     ax1.set_ylabel('Probability Density')
31     ax2.legend(title="normal distributions")
32     ax2.set_xlabel('ln(x)')
33     ax2.set_ylabel('Probability Density')
34
35     fig.tight_layout()
```

Listing 9.4 Generating random samples with log-normal distributions

Note that I am using log and log10 for the natural and base 10 logarithms, respectively. This is the same notation we find in Python. We can generate a log-normal distribution using the *scipy.stats.lognorm()* method. Among the others, it accepts *s* as the shape parameter. Also, *scale*, and *loc* enable you to shift and scale the distribution.[1] To generate a log-normal distribution as defined in Eq. (9.5), *s* and *scale* must be set to σ_n and e^{μ_n}, respectively (see code Listing 9.4 and Fig. 9.3).

9.4   Other Useful PDFs for Geological Applications

The scipy.stats module allows the management of many probability distributions, that are useful in geological applications. Examples include the Poisson, Pareto, and the Student's t distributions, which find use in fields such as geochemical determinations [82], metal exploration [61], and geophysical investigations [81]. Table 9.1 reports various probability distributions from the scipy.stats module.

9.5   Density Estimation

Estimating density consists of reconstructing PDFs from the observed data [66, 77]. I describe two main approaches to achieve this goal: The first approach is parametric and consists of fitting the observed data using a known PDF [66, 77]. For example, to fit a bell-shaped distribution with a normal PDF, we start by estimating its mean μ and standard deviation σ. Next, the obtained μ and σ are used to reconstruct a normal PDF and fit the observed distribution. The processes of fitting described in code Listings 9.3 and 9.4 are all examples of parametric density estimations.

---

[1] https://docs.scipy.org/doc/scipy/reference/generated/scipy.stats.lognorm.html.

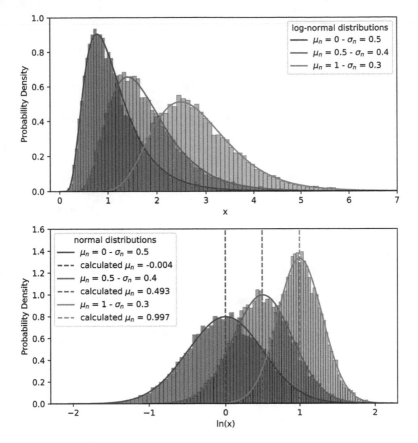

Fig. 9.3 Result of code Listing 9.4

Although intriguing for its simplicity, the parametric approach is not always the best choice [66, 77]. For example, popular PDFs are mostly unimodal, but many practical examples in geology involve multimodal distributions. Also, choosing a specific known PDF is not always straightforward when working with real geological applications. Consequently, the so-called non-parametric approach is often the best choice [66, 77]: it estimates the density directly from the data, without making any parametric assumptions about the underlying distribution [66, 77].

A density histogram is the simplest form of a non-parametric density estimation [66, 77]. We encountered density histograms earlier in this book in Sect. 4.2. The development of a density histogram is quite easy: it consists of dividing the the sample space into intervals called bins [66, 77] and then estimating the density of each bin by using [66, 77]

$$\hat{f}(x_i - h/2 \le x < x_i + h/2) = \frac{k_i}{nh}, \tag{9.6}$$

Table 9.1 Selected statistical functions from the scipy.stats module

Function	Distribution	Function	Distribution
alpha()	Alpha cont. random var.	arcsine()	Arcsine cont. random var.
beta()	Beta cont. random var.	bradford()	Bradford cont. random var.
cauchy()	Cauchy cont. random var.	chi()	Chi cont. random var.
chi2()	Chi-squared cont. random var.	cosine()	Cosine cont. random var.
dgamma()	Double gamma cont. random var.	dweibull()	Double Weibul cont. random var.
erlang()	Erlang cont. random var.	expon()	Exponential cont. random var.
halfcauchy()	Half-Cauchy cont. random var.	halfnorm()	Half-normal cont. random var.
laplace()	Laplace cont. random var.	levy()	Levy cont. random var.
logistic()	Logistic cont. random var.	loggamma()	Log gamma cont. random var.
loglaplace()	Log-Laplace cont. random var.	loguniform()	Loguniform cont. random var.
maxwell()	Maxwell cont. random var.	pareto()	Pareto cont. random var.
pearson3()	Pearson type III cont. random var.	powerlaw()	Power-function cont. random var.
rayleigh()	Rayleigh cont. random var.	skewnorm()	Skew-normal cont. random var.
t()	Student's t cont. random var.	uniform()	Uniform cont. random var.
bernoulli()	Bernoulli discr. random var.	binom()	Binomial discr. random var.
boltzmann()	Boltzmann discr. random var.	dlaplace()	Laplacian discr. random var.
geom()	Geometric discr. random var.	poisson()	Poisson discr. random var.
gamma()	A gamma cont. random var.	pareto()	A Pareto cont. random var.

where x_i is the x value at the center of each bin, (i.e., the interval $[x_i - h/2, x_i + h/2]$), k_i is the number of observations in the interval $x_i - h/2 \leq x_i < x_i + h/2$, n is the number of bins, and h is the bin width (i.e., $h = x_i - x_{i+1} = (x_{max} - x_{min})/n$. Note that the symbol $\hat{f}$ refers to the empirical estimate of the PDF.

```python
1  from statsmodels.nonparametric.kde import KDEUnivariate
2  import matplotlib.pyplot as plt
3  import numpy as np
4
5  kernels = ['gau', 'epa', 'uni', 'tri', 'biw', 'triw']
6  kernels_names = ['Gaussian', 'Epanechnikov', 'Uniform', '
       Triangular', 'Biweight', 'Triweight']
7  positions = np.arange(1,9,1)
8
9  fig, ax = plt.subplots()
10
11 for kernel, kernel_name, pos in zip(kernels, kernels_names,
       positions):
12
13     # kernels
14     kde = KDEUnivariate([0])
15     kde.fit(kernel= kernel, bw=1, fft=False, gridsize=2**10)
16     ax.plot(kde.support, kde.density, label = kernel_name,
       linewidth=1.5, alpha=0.8)
17
18 ax.set_xlim(-2,2)
19 ax.grid()
20 ax.legend(title='kernel functions')
```

Listing 9.5 Kernel functions available in *KDEUnivariate()*

To estimate a PDF starting from experimental data, a more evolved method than the density histograms is the kernel density estimation (KDE). A KDE is a non-parametric way to estimate the PDF of a random variable. To understand, let $(x_1, x_2, x_i, \ldots, x_n)$ be a univariate, independent, and identically distributed sample (i.e., all x_i have the same probability distribution) belonging to a distribution with an unknown PDF. We are interested in estimating the shape $\hat{f}$ of this PDF. The equation defining a KDE is

$$\hat{f}(x) = \frac{1}{nh} \sum_{i=1}^{n} K\left(\frac{x - x(i)}{h}\right), \qquad (9.7)$$

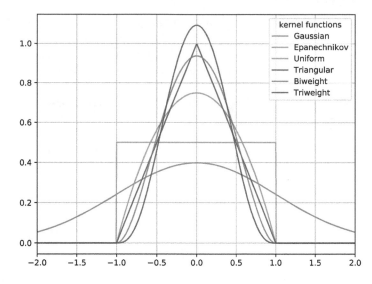

Fig. 9.4 Result of code Listing 9.5

Table 9.2 Selection of kernel density estimators in Python

Package	Function	Description
Scipy	gaussian_kde()	Kernel-density estimate using Gaussian kernels
Statsmodels	KDEUnivariate()	Univariate kernel density estimator
Statsmodels	KDEMultivariate()	Multivariate kernel density estimator
Scikit-Learn	KernelDensity()	Multivariate kernel density estimator
Seaborn	kdeplot()	Plot univariate or bivariate distributions using kernel density estimation

where K is the kernel, a non-negative function that integrates to unity [i.e., $\int_{-\infty}^{\infty} K(x)dx = 1$], and $h > 0$ is a smoothing parameter called the bandwidth. A range of kernel functions are commonly used: normal, uniform, triangular, biweight, triweight, Epanechnikov, and others (code Listing 9.5 and Fig. 9.4).

Python offers many different implementations for the development of a KDE (Table 9.2).

```
1  from statsmodels.nonparametric.kde import KDEUnivariate
2  import pandas as pd
3  import numpy as np
4  import matplotlib.pyplot as plt
5
6  my_dataset = pd.read_excel('Smith_glass_post_NYT_data.xlsx',
       sheet_name='Supp_traces')
7
8  x = my_dataset.Zr
9  x_eval = np.arange(0, 1100, 1)
10
11 fig = plt.figure()
12
13 ax1 = fig.add_subplot(2, 1, 1)
14 # Density Histogram
15 ax1.hist(x, bins='auto', density=True, label='Density Histogram
       ', color='#c7ddf4', edgecolor='#000000')
16 kde = KDEUnivariate(x)
17 kde.fit()
18 my_kde = kde.evaluate(x_eval)
19 ax1.plot(x_eval, my_kde, linewidth=1.5, color='#ff464a', label=
       'gaussian KDE - auto bandwidth selection')
20 ax1.set_xlabel('Zr [ppm]')
21 ax1.set_ylabel('Probability density')
22 ax1.legend()
23
24 ax2 = fig.add_subplot(2, 1, 2)
25 # Density Histogram
26 ax2.hist(x, bins= "auto", density = True, label='Density
       Histogram', color='#c7ddf4', edgecolor='#000000')
27
28 # KDE
29 # Effect of bandwidth
30 for my_bw in [10,50,100]:
31
32     kde = KDEUnivariate(x)
33     kde.fit(bw = my_bw)
34
35     my_kde = kde.evaluate(x_eval)
36     ax2.plot(x_eval, my_kde, linewidth = 1.5, label='gaussian
       KDE - bw: ' + str(my_bw))
37
38 ax2.set_xlabel('Zr [ppm]')
39 ax2.set_ylabel('Probability density')
40 ax2.legend()
41
42 fig.tight_layout()
```

Listing 9.6 Example application of KDE in geology: effect of bandwidth

```
1   import pandas as pd
2   import matplotlib.pyplot as plt
3   import numpy as np
4   from statsmodels.nonparametric.kde import KDEUnivariate
5
6   # import Zircon data from Puetz (2010)
7   my_data = pd.read_excel('1-s2.0-S1674987117302141-mmc1.xlsx',
        sheet_name='Data')
8   my_data = my_data[(my_data.age206Pb_238U>0)&(my_data.
        age206Pb_238U<1500)]
9   my_sample = my_data.age206Pb_238U
10
11  # Plot the Density Histogram
12  fig, ax = plt.subplots(figsize=(8,5))
13  bins = np.arange(0,1500,20)
14  ax.hist(my_sample, bins, color='#c7ddf4', edgecolor='k',
        density=True, label='Density Histogram - bins = 20 My')
15
16  # Compute and plot the KDE
17  age_eval = np.arange(0,1500,10)
18  kde = KDEUnivariate(my_sample)
19  kde.fit(bw=20)
20  pdf = kde.evaluate(age_eval)
21  ax.plot(age_eval, pdf, label ='Gaussian KDE - bw = 20 Ma',
        linewidth=2, alpha=0.7, color='#ff464a')
22
23  # Adjust diagram parameters
24  ax.set_ylim(0,0.0018)
25  ax.set_xlabel('Age (My)')
26  ax.set_ylabel('Probability Densisty')
27  ax.legend()
28  ax.grid(axis='y')
29
30  # Plot mass extinction annotations
31  mass_extinction_age = [444, 359, 252, 66, 0]
32  pdf_mass_extinction_age = kde.evaluate(mass_extinction_age)
33  mass_extincyion_name = ["Ordovician-Silurian", "Late Devonian",
        "Permian-Triassic", "Cretaceous-Paleogene", "Triggered by
        Men?"]
34  y_offsets = [0.0001, 0.0001, 0.0002, 0.0002, 0.0004]
35  y_texts = [30, 105, 15, 62, 160]
36  x_texts = [30, 30, 30, 30, 30]
37
38  for x, y, name, x_text, y_text, y_offset in zip(
        mass_extinction_age, pdf_mass_extinction_age,
        mass_extincyion_name, x_texts, y_texts, y_offsets):
39      ax.annotate(name, xy=(x, y + y_offset), xycoords='data',
40      xytext=(x_text, y_text), textcoords='offset points',
41      arrowprops=dict(arrowstyle="->",
42      connectionstyle="angle, angleA=0, angleB=90, rad=10"))
43
44  fig.tight_layout()
```

Listing 9.7 Example application of KDE in geology

Code Listing 9.6 and Fig. 9.5 show the application of the *KDEUnivariate()* function to geochemical data and how the bandwidth affects the resulting KDE estimate.

As an example application of density histograms and KDEs for unravelling PDFs in geological applications, the code Listing 9.7 and Fig. 9.6 show the reconstruction

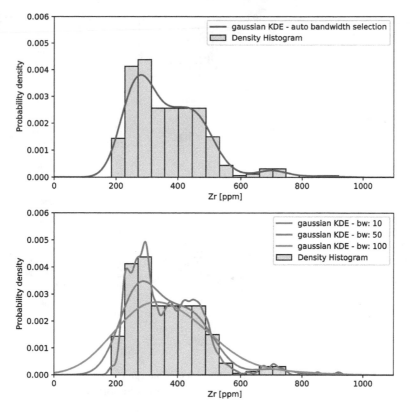

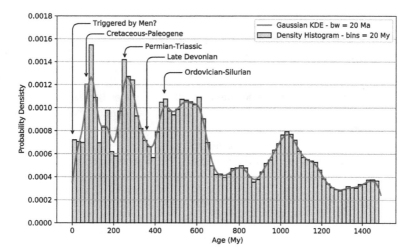

Fig. 9.5 Result of code Listing 9.6

Fig. 9.6 Result of code Listing 9.7

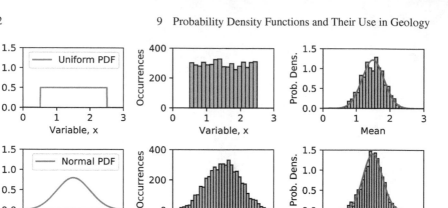

Fig. 9.7 Result of code Listing 9.8

of ^{238}U/^{206}Pb zircon ages for the last 1500 My. The data are from [72]. Due to the recent re-rising linking magmatism to mass extinction [70, 79], the largest extinction events are also reported.

9.6   The Central Limit Theorem and Normal Distributed Means

The Central Limit Theorem can be stated in different ways, the easiest of which is that given by [68]: "the sum of a large number of independent random variables, each with finite mean and variance, will tend to be normally distributed, irrespective of the distribution function of the random variable."

To familiarize ourselves with the Central Limit Theorem, code Listing 9.8 and Fig. 9.7 replicate the experiment reported in Fig. 3.7 by Hughes and Hase [68].

```
1  import numpy as np
2  import scipy.stats as stats
3  import matplotlib.pyplot as plt
4
5  fig = plt.figure(figsize=(8,6))
6
7  dists = [stats.uniform(loc=0.5, scale=2), stats.norm(loc=1.5,
          scale=0.5), stats.laplace(loc=1.5, scale=0.6)]
8  names = ['Uniform', 'Normal', 'Laplace']
9  x = np.linspace(0,3,1000)
10
11 for i, (dist, name) in enumerate(zip(dists, names)):
12
13     # Probability Density Function (pdf)
14     pdf = dist.pdf(x)
15     ax1 = fig.add_subplot(3, 3, 3*i+1)
16     ax1.plot(x, pdf, color='#4881e9', label= name + ' PDF')
17     ax1.set_xlim(0,3)
18     ax1.set_ylim(0,1.5)
19     ax1.set_xlabel('Variable, x')
20     ax1.set_ylabel('Prob. Dens.')
21     ax1.legend()
22
23     #Distribution (rnd) of the Random Variable based on the
          selected pdf
24     rnd = dist.rvs(size=5000)
25     ax2 = fig.add_subplot(3, 3, 3*i+2)
26     ax2.hist(rnd, bins='auto', color='#84b4e8', edgecolor='
          #000000')
27     ax2.set_xlim(0,3)
28     ax2.set_ylim(0,400)
29     ax2.set_xlabel('Variable, x')
30     ax2.set_ylabel('Occurrences')
31
32     ax3 = fig.add_subplot(3, 3, 3*i+3)
33     mean_dist = []
34     for _ in range(1000):
35         mean_dist.append(dist.rvs(size=3).mean())
36     mean_dist = np.array(mean_dist)
37     ax3.hist(mean_dist, density=True, bins='auto', color='#84
          b4e8', edgecolor='#000000')
38     normal = stats.norm(loc= mean_dist.mean(), scale= mean_dist
          .std())
39     ax3.plot(x, normal.pdf(x), color='#ff464a')
40     ax3.set_xlim(0,3)
41     ax3.set_ylim(0,1.5)
42     ax3.set_xlabel('Mean')
43     ax3.set_ylabel('Prob. Dens.')
44
45 fig.tight_layout()
```

Listing 9.8 The Central Limit Theorem [68]

In detail, code Listing 9.8 starts from three different distributions of the random variable (i.e., uniform, normal, and Laplace; see Table 9.1) to create (1) the relative PDF (first column of Fig. 9.7), (2) 1000 randomly generated occurrences of the random variable (second column of Fig. 9.7), and (3) the estimate of the mean value of the distribution based on 1000 attempts using three randomly selected occurrences of the random variable (third column of Fig. 9.7).

In accordance with the Central Limit Theorem, the histograms of the estimated means are normally distributed (third column of Fig. 9.7) with a peak mean at 1.5. Also, the distribution of the means (third column of Fig. 9.7) is narrower than the original distributions (second column of Fig. 9.7) by a factor $\sqrt{N}$. Further details and some geological implications of the Central Limit Theorem are presented and discussed in Chap. 10.

Chapter 10
Error Analysis

10.1 Dealing with Errors in Geological Measurements

As reported by Hughes and Hase [68], the aim of error analysis is to quantify and record the errors associated with the inevitable spread in a set of measurements. This is also true for geological estimates. The following definitions are taken from the book "Measurements and their Uncertainties" by Hughes and Hase [68]. Two fundamental terms describe the uncertainties associated with a set of measurements: precision and accuracy. An accurate measurement is one in which the results of the experiments are consistent with the accepted value. A precise result is one where the spread of measurements is "small" either relative to the average results or in absolute magnitude. This chapter also discusses the standard error (i.e., the uncertainty in estimates of the mean) and how to propagate uncertainties by using either the linear method or the Monte Carlo approach.

Precision and Accuracy

To introduce the concepts of precision and accuracy, I use a practical example: the estimate of the figure of merit of an instrument used to chemically characterize geological samples. The definition of the precision and the accuracy of an analytical device is typically obtained by using a reference material such as a homogeneous chemical sample of known composition (better if certified), analyzed as an unknown. The following results were obtained during repeated analyses of the USGS BCR2G reference material at the LA-ICP-MS facility of Perugia University over about five years. These results were obtained in very comfortable operating conditions using a large beam diameter (80 μm), a frequency of 10 Hz, and a laser fluence of $\approx$3.5 J/cm^2. The chemical element reported here is lanthanum (La), which is present at a concentration of 25.6 $\pm$ 0.5 ppm (Rocholl [74]). Data are stored in the USGS_BCR2G.xls file.

© The Author(s), under exclusive license to Springer Nature Switzerland AG 2021   155
M. Petrelli, *Introduction to Python in Earth Science Data Analysis*,
Springer Textbooks in Earth Sciences, Geography and Environment,
https://doi.org/10.1007/978-3-030-78055-5_10

```
1  import pandas as pd
2  import scipy.stats as stats
3  import matplotlib.pyplot as plt
4  import numpy as np
5
6  my_dataset = pd.read_excel('USGS_BCR2G.xls', sheet_name='Sheet1')
7
8  fig, ax = plt.subplots()
9  ax.hist(my_dataset.La, bins='auto', density=True, edgecolor='
        #000000', color='#c7ddf4', label="USGS BCR2G")
10 ax.set_xlabel("La [ppm]")
11 ax.set_ylabel("Probability Density")
12
13 x = np.linspace(23,27.5,500)
14 pdf = stats.norm(loc=my_dataset.La.mean(), scale=my_dataset.La.std
        ()).pdf(x)
15
16 ax.plot(x,pdf, linewidth=2, color='#ff464a', label='Normal
        Distribution')
17
18 ax.legend()
```

Listing 10.1 LA-ICP-MS determinations of La in the USGS BCR2G reference material

Specifically, accuracy measures the agreement of our estimates with real values. Typically, the accuracy of an analytical device (LA-ICP-MS in our case) is estimated by evaluating the agreement between the estimates and the accepted values of a reference material. The deviation of the mean μ of the measurements from the accepted value R is an estimate of accuracy:

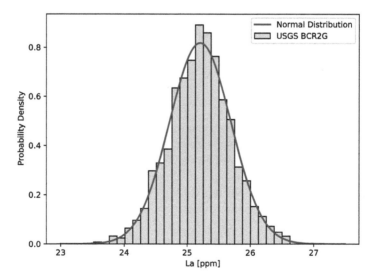

Fig. 10.1 Result of code Listing 10.1

$$\text{Accuracy } [\%] = \frac{\mu - R}{R} \times 100. \tag{10.1}$$

The precision of a set of measurements is the spread in the distribution of measured values and can be estimated by using an index of dispersion (Chap. 5). Typically, the standard deviation serves as the metric of dispersion and is often expressed in percent:

$$\text{Precision } [\%] = \frac{\sigma}{R} \times 100. \tag{10.2}$$

```
1  my_mean = my_dataset.La.mean()
2  R = 25.6
3  accuracy = 100 * (my_mean - R) / R
4  my_std = my_dataset.La.std()
5  precision = 100 * my_std / R
6
7  fig, ax = plt.subplots(figsize=(6,5))
8  ax.hist(my_dataset.La, bins = 'auto', density = True, edgecolor =
       '#000000', color = '#c7ddf4', label = 'USGS BCR2G')
9  ax.set_xlabel('La [ppm]')
10 ax.set_ylabel('Probability Density')
11
12 ax.axvline(x=my_dataset.La.mean(), color='#ff464a', linewidth=3,
       label='Mean of the Measurements:' + str(round(my_mean, 1)) + '
       [ppm]')
13 ax.axvline(x = R, color='#342a77', linewidth=3, label='Accepded
       Value')
14
15 ax.axvline(x = my_mean - my_std, color = '#4881e9', linewidth = 1)
16 ax.axvline(x = my_mean + my_std, color = '#4881e9', linewidth = 1)
17 ax.axvspan(my_mean - my_std,  my_mean + my_std, alpha = 0.2, color
       = '#342a77', label = r'$1\sigma$')
18 ax.legend(loc='upper center', bbox_to_anchor=(0.5, -0.15),
       fancybox=False, shadow=False, ncol=2, title = 'Accuracy = {:.1
       f} % - Precision = {:.1f} %'.format(accuracy, precision))
19
20 fig.tight_layout()
```
Listing 10.2 Accuracy and precision

Confidence Intervals

As a consequence of the Central Limit Theorem, a sufficiently large set of measurements of the same target will approach a normal distribution due to the many random sources of (small) uncertainty (Fig. 10.1; see Sect. 9.6 for further details) (Fig. 10.2).

Therefore, by fitting the histogram diagram reported in Fig. 10.1 to a normal distribution, we can highlight the probability of La measurements to lie within one (68.27%), two (95.45%), or three standard deviations (99.27%) around the mean value (Eq. (9.4), code Listing 10.3 and Fig. 10.3). To note, Fig. 10.3 simply shows the rationale of providing an estimate for a quantity x by using the mean value μ_x and the confidence intervals [68, 78]:

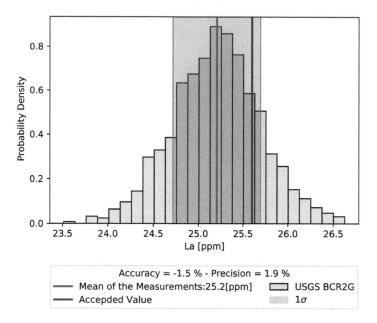

Fig. 10.2 Result of code Listing 10.2

$$\mu \pm n\sigma_x, \qquad (10.3)$$

with $n = 1, 2, 3, \ldots$ corresponding to confidence intervals of 68.27%, 95.45%, 99.27%, ..., respectively.

```
1  import numpy as np
2  import matplotlib.pyplot as plt
3
4
5  def normal_pdf(x, mu, sigma):
6      pdf = 1/(sigma*np.sqrt(2*np.pi)) * np.exp(-(x-mu)**2 / (2*
       sigma**2))
7      return pdf
8
9  signa_levels = [1, 2, 3]
10 confidences = [68.27, 95.45, 99.73]
11
12 fig = plt.figure(figsize=(7,8))
13
14 my_mean = my_dataset.La.mean()
15 my_std = my_dataset.La.std()
16
17 x_pdf = np.linspace(my_mean - 4 * my_std, my_mean + 4 * my_std,
       1000)
18 my_pdf = normal_pdf(x_pdf, my_mean, my_std)
19
20 for signa_level, confidence in zip(signa_levels,confidences):
21     ax = fig.add_subplot(3, 1, signa_level)
```

```
22    ax.hist(my_dataset.La, bins='auto', density=True, edgecolor='
        #000000', color='#c7ddf4', label='USGS BCR2G', zorder=0)
23    x_confidence = np.linspace(my_mean - signa_level * my_std,
        my_mean + signa_level * my_std, 1000)
24    ax.plot(x_pdf, my_pdf, linewidth=2, color='#ff464a', label='
        Normal Distribution', zorder=1)
25    ax.fill_between(x_confidence, normal_pdf(x_confidence, my_mean
        , my_std), y2=0, color='#ff464a', alpha=0.2, label='prob. = {}
        '.format(confidence) + ' %', zorder=1)
26    ax.legend(ncol=3, loc='upper center', title=r'$\mu~ \pm ~$' +
        str(signa_level) + r'$ ~ \sigma ~ $ = ' + '{:.1f}'.format(
        my_mean) + r'$~ \pm ~$' + '{:.1f}'.format(signa_level * my_std
        ))
27    ax.set_ylim(0,1.6)
28    ax.set_xlabel('La [ppm]')
29    ax.set_ylabel('prob. dens.')
30
31  fig.tight_layout()
```

Listing 10.3 Confidence intervals

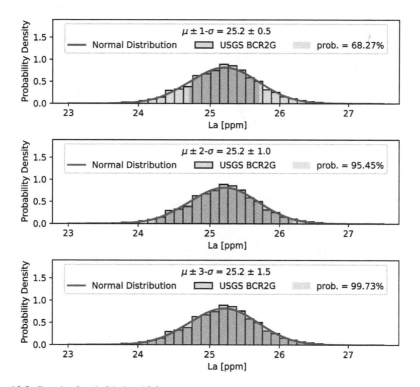

Fig. 10.3 Result of code Listing 10.3

Uncertainties of Mean Estimates: Standard Error

The standard deviation of the mean, or standard error SE, is a measure of the uncertainty in the location of the mean of a set of measurements [68]:

$$SE = \frac{\sigma_s}{\sqrt{n}}. \tag{10.4}$$

Consequently, mean estimates μ_s should be reported as [68, 78]

$$\mu_s \pm SE = \mu_s \pm \frac{\sigma_s}{\sqrt{n}}. \tag{10.5}$$

The significance of Eq. (10.4) can be evaluated in light of the Central Limit Theorem. Assume you are sampling a homogeneous material (e.g., a geological reference material like the USGS BCR2G) characterized by a perfectly known target value of 1.5 (the units are not important here) using a well-calibrated analytical device (i.e., no accuracy biases).

```
1   import numpy as np
2   import scipy.stats as stats
3   import matplotlib.pyplot as plt
4
5   mean_value = 1.5
6   std_dev = 0.5
7   dist = stats.norm(loc=mean_value, scale=std_dev)
8   x = np.linspace(0, 3, 1000)
9   fig = plt.figure(figsize=(6,8))
10
11  # Distribution of the Random Variable based on the normal PDF
12  pdf = dist.pdf(x)
13  ax1 = fig.add_subplot(3, 1, 1)
14  ax1.plot(x, pdf, color='#84b4e8', label =r'$\mu_p$ = 1.5 - 1$\
        sigma_p$ = 0.5')
15  ax1.set_xlim(0,3)
16  ax1.set_ylim(0,1)
17  ax1.set_xlabel('Variable, x')
18  ax1.set_ylabel('Prob. Dens.')
19  ax1.legend(title = 'Parent Distribution')
20
21  # Dependence of the SE on the Central Limit Theorem
22  ax2 = fig.add_subplot(3, 1, 2)
23  std_of_the_mean = []
24  ns = [2, 10, 100, 500]
25
26  for n in ns:
27      # Mean Estimation Based on 1000 attempts using N values
28      mean_dist = []
29      for _ in range(1000):
30          mean_dist.append(dist.rvs(size=n).mean())
31      mean_dist = np.array(mean_dist)
32      std_of_the_mean.append(mean_dist.std())
33      normal = stats.norm(loc=mean_dist.mean(), scale=mean_dist.std
          ())
34      ax2.plot(x, normal.pdf(x), label='N = ' + str(n))
35  ax2.set_xlim(0, 3)
```

```
36  ax2.set_xlabel('Mean')
37  ax2.set_ylabel('Prob. Dens.')
38  ax2.legend(title='Standard Deviation of the Means', ncol=2)
39
40  # SE estimates and the empirically derived std of the Means
41  ax3 = fig.add_subplot(3, 1, 3)
42  ax3.scatter(ns, std_of_the_mean, color='#ff464a', edgecolor='
        #000000', label='Standard Deviation of the Means', zorder = 1)
43  n1 = np.linspace(1, 600, 600)
44  se = std_dev / np.sqrt(n1)
45  ax3.plot(n1 , se, c='#4881e9', label='Standard Error (SE)', zorder
        =0)
46  ax3.set_xlabel('N')
47  ax3.set_ylabel('Standard Error, SE')
48  ax3.legend()
49  fig.tight_layout()
```

Listing 10.4 Standard error estimate

Given the numerous random uncertainties associated with the analytical device, the target population (i.e., the set of all possible measurements) will assume a normal distribution, in agreement with the Central Limit Theorem (see upper panel of Fig. 10.4). To make the analysis, we start sampling the target population. What is the uncertainty associated with the mean estimate based on n estimates? The standard error is a measure of this uncertainty and is measured either by using Eq. (10.4) or by repeating the mean estimate many times (1000 in the case of code Listing 10.4) with N measurements and estimating the standard deviation of the set of means obtained (see middle panel of Fig. 10.4). Being geologists, we only trust the evidence, so the bottom panel of Fig. 10.4 compares the SE obtained using Eq. (10.4) with the distribution of the standard deviation of the mean obtained in the above experiment. Code Listing 10.4 shows the procedure to unravel the meaning of the SE and create Fig. 10.4.

But what information is provided by the SE? To answer this, consider code Listing 10.4 and Fig. 10.4, where we are sampling (e.g., analyzing an unknown geological material, or sampling a geological quantity such as the dip angle or the strike of bedding) the same normal population of Fig. 10.4 characterized by a mean and standard deviation of 1.5 and 0.5, respectively. Making only three estimates gives mean and standard-deviation estimates of 1.56 and 0.51, respectively. In this case, $SE = 0.23$, so we should write $\mu_s = 1.56 \pm 0.23$ and $\sigma_s = 0.51$. Note that three parameters are required to define our measurements. Upon increasing n, SE decreases progressively, with μ_s becoming a more robust estimate of the mean value of the parent distribution (Fig. 10.4).

Always remember that the standard deviation is a measure of the spread of the sampled distribution. It highlights how accurately the mean represents the sampled distribution. In contrast, the standard error measures how far the sample mean μ_s of the measurements is likely to be from the true population mean μ_p. Finally, note that SE is always less than σ_s.

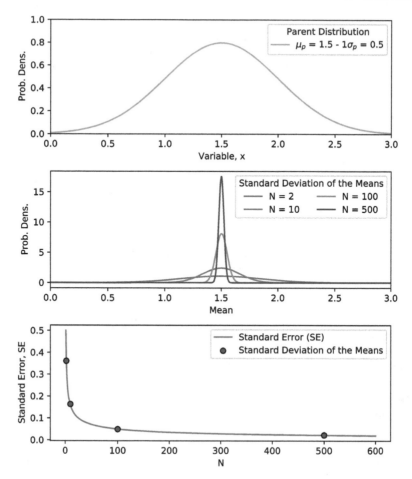

Fig. 10.4 Result of code Listing 10.4

10.2   Reporting Uncertainties in Binary Diagrams

Errors are always present in empirical estimates (e.g., geological samplings and analytical determinations). Consequently, uncertainties should always be taken into account during data visualization and modeling. Assuming a normal distribution of our estimates (cf. the Central Limit Theorem described in Sect. 9.6), we can set confidence levels at 68%, 95%, and 99.7% using 1σ, 2σ, and 3σ, respectively.

```
1  import pandas as pd
2  import matplotlib.pyplot as plt
3
4  my_dataset1 = pd.read_excel('Smith_glass_post_NYT_data.xlsx',
       sheet_name='Supp_traces')
5
```

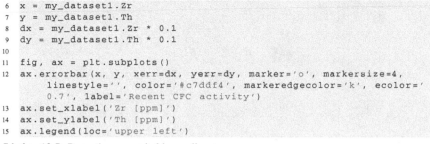

```
 6  x = my_dataset1.Zr
 7  y = my_dataset1.Th
 8  dx = my_dataset1.Zr * 0.1
 9  dy = my_dataset1.Th * 0.1
10
11  fig, ax = plt.subplots()
12  ax.errorbar(x, y, xerr=dx, yerr=dy, marker='o', markersize=4,
        linestyle='', color='#c7ddf4', markeredgecolor='k', ecolor='
        0.7', label='Recent CFC activity')
13  ax.set_xlabel('Zr [ppm]')
14  ax.set_ylabel('Th [ppm]')
15  ax.legend(loc='upper left')
```

Listing 10.5 Reporting errors in binary diagrams

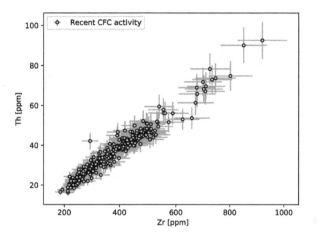

Fig. 10.5 Result of code Listing 10.5

```
 1  import numpy as np
 2  import matplotlib.pyplot as plt
 3
 4  x = np.array([250,300,360,480,570,770,870,950])
 5  y = np.array([20,25,30,40,50,70,80,100])
 6
 7
 8  fig = plt.figure(figsize=(6,8))
 9
10  # xerr and yerr reported as single value
11  dx = 50
12  dy = 10
13  ax1 = fig.add_subplot(3,1,1)
14  ax1.errorbar(x, y, xerr=dx, yerr=dy, marker='o', markersize=6,
        linestyle = '', color='#c7ddf4', markeredgecolor='k', ecolor='
        0.7', label='single value for xerr and yerr')
15  ax1.legend(loc='upper left')
16
17  # xerr and yerr reported as 1D array
18  dx = np.array([25,35,40,120,150,30,30,25])
19  dy = np.array([8,8,6,7,7,35,40,40])
20
21  ax2 = fig.add_subplot(3,1,2)
22  ax2.errorbar(x, y, xerr=dx, yerr=dy, marker='o', markersize=6,
        linestyle = '', color='#c7ddf4', markeredgecolor='k', ecolor='
        0.7', label='xerr and yerr as 1D array')
23  ax2.set_ylabel('Th [ppm]')
24  ax2.legend(loc='upper left')
25
26  # xerr and yerr reported as 2D array
27  dx = np.array
        ([[80,60,70,100,150,150,20,100],[20,25,30,30,30,30,90,30]])
28  dy = np.array([[10,4,10,15,15,20,5,5],[2,8,4,4,6,7,10,20]])
29
30  ax3 = fig.add_subplot(3,1,3)
31  ax3.errorbar(x, y, xerr=dx, yerr=dy, marker='o', markersize=6,
        linestyle = '', color='#c7ddf4', markeredgecolor='k', ecolor='
        0.7',  label='xerr and yerr as 2D array')
32  ax3.set_xlabel('Zr [ppm]')
33  ax3.legend(loc='upper left')
34
35  fig.tight_layout()
```

Listing 10.6 Reporting errors in binary diagrams

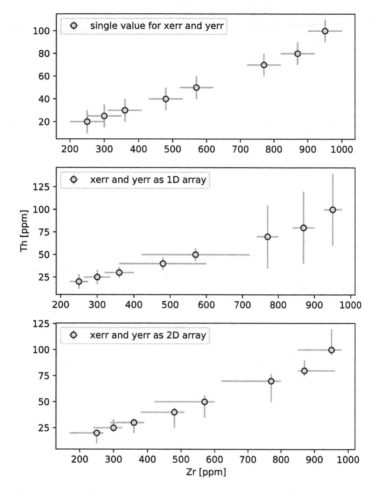

Fig. 10.6 Result of code Listing 10.6

In binary diagrams, errors can easily be reported by using the *errorbar()* function of the matplotlib.pyplot sub-package (code Listing 10.5 and Fig. 10.5).

The *errorbar()* function accepts all arguments available to *plot()*, plus *xerr*, *yerr*, and the related arguments. *xerr* and *yerr* refer to the error on the *x* and *y* axes, respectively.

```
1  import numpy as np
2  import matplotlib.pyplot as plt
3
4  x = np.array([200, 300, 360, 480, 570, 770, 870, 950])
5  y = np.array([10, 15, 30, 40, 50, 70, 80, 100])
6  dx = 40
7  dy = 10
8
9  fig = plt.figure()
10 ax1 = fig.add_subplot(2, 1, 1)
```

```
11  ax1.errorbar(x, y, xerr=dx, yerr=dy, marker='o', markersize=4,
        linestyle='', color='k',  ecolor='0.7', elinewidth=3, capsize
        =0, label='Recent activity of the CFC')
12  ax1.legend(loc='upper left')
13  ax1.set_xlabel('Zr [ppm]')
14  ax1.set_ylabel('Th [ppm]')
15
16  ax2 = fig.add_subplot(2,1,2)
17  ax2.errorbar(x, y, xerr=dx, yerr=dy, marker='o', markersize=6,
        linestyle='', color='#c7ddf4', markeredgecolor='k', ecolor='k'
        , elinewidth = 0.8, capthick=0.8, capsize=3, label='Recent
        activity of the CFC')
18  ax2.legend(loc='upper left')
19  ax2.set_xlabel('Zr [ppm]')
20  ax2.set_ylabel('Th [ppm]')
```

Listing 10.7 Reporting errors in binary diagrams

Also, they can be a one- or two-dimensional arrays. Using one-dimensional arrays (e.g., Fig. 10.5), a symmetrical error (i.e., $x \pm xerr$) is defined for each single point. Finally, by reporting *xerr* and *yerr* as a two-dimensional array, we can report non-symmetrical errors (see code Listing 10.6 and Fig. 10.6). Finally, to provide publication-ready diagrams, error reporting can be personalized in many different ways (e.g., Figs. 10.7 and 10.8).

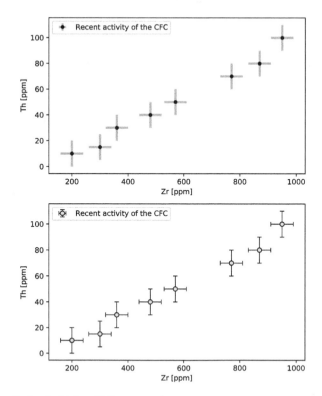

Fig. 10.7 Result of code Listing 10.7

```
1  import pandas as pd
2  import matplotlib.pyplot as plt
3
4
5  def plot_errorbar(x,y, dx, dy, xoffset, yoffset, text, ax):
6      ax.errorbar(x,y, xerr=dx, yerr=dy, marker='', linestyle = '',
          elinewidth = .5, capthick=0.5, ecolor='k', capsize=3)
7      ax.text(x + xoffset, y + yoffset, text)
8
9  my_dataset1 = pd.read_excel('Smith_glass_post_NYT_data.xlsx',
          sheet_name='Supp_traces')
10
11  x = my_dataset1.Zr
12  y = my_dataset1.Th
13
14  dx = 60
15  dy = 7
16
17  errorbar_x = x.max() - x.max() * 0.1
18  errorbar_y = y.min() + y.max() * 0.1
19
20  fig, ax1 = plt.subplots()
21  ax1.scatter(x, y, marker='o', color='#4881e9', edgecolor='k',
          alpha=0.8, label='Recent activity of the CFC')
22
23  plot_errorbar(errorbar_x, errorbar_y, dx, dy, dx/4, dy/4, r'2$\
          sigma$', ax1)
24
25  ax1.legend(loc='upper left')
26  ax1.set_xlabel('Zr [ppm]')
27  ax1.set_ylabel('Th [ppm]')
```

Listing 10.8 Reporting errors in binary diagrams

Fig. 10.8 Result of code
Listing 10.8

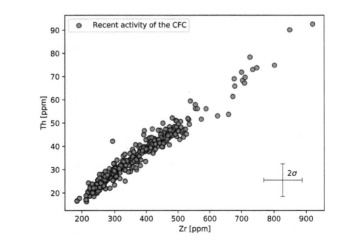

10.3   Linearized Approach to Error Propagation

When using a linearized approximation (i.e., a first-order Taylor series expansion) and assuming uncorrelated and statistically independent variables (i.e., the independent variables are uncorrelated with the magnitude and the error of all other parameters), the general formula for error propagation takes the form [68, 78])

$$\sigma_z = \sqrt{\left(\frac{\partial z}{\partial a}\right)^2 (\sigma_a)^2 + \left(\frac{\partial z}{\partial b}\right)^2 (\sigma_b)^2 + \left(\frac{\partial z}{\partial c}\right)^2 (\sigma_c)^2 + \cdots}, \qquad (10.6)$$

where z is a multi-variable function $z = f(a, b, c, \dots)$ that depends on the measured variables $a \pm \sigma_a$, $b \pm \sigma_b$, $c \pm \sigma_c$, etc. Table 10.1 shows the result of applying Eq. (10.6) to some simple, common equations that are often useful to solve geological problems. If correlations between the variables involved cannot be neglected (i.e., they are not independent), additional terms should be added. For example, given the function $z = f(x, y)$, which depends on measured quantities $x \pm \sigma_x$ and $y \pm \sigma_y$, with covariance σ_{xy} between x and y, the uncertainty in z is

$$\sigma_z = \sqrt{\left(\frac{\partial z}{\partial x}\right)^2 (\sigma_x)^2 + \left(\frac{\partial z}{\partial y}\right)^2 (\sigma_y)^2 + 2\frac{\partial z}{\partial x}\frac{\partial z}{\partial y}\sigma_{xy}}. \qquad (10.7)$$

Keep in mind that the reported linearized approach, based on the first-order Taylor series expansion, assumes that the magnitude of the error is small [68, 78]. Conse-

Table 10.1 Error propagation for common equations that are often useful to solve geological problems. Modified from Hughes and Hase [68]

Function z	Error	Function z	Error
$z = 1/a$	$\sigma_z = z^2 \sigma_a$	$z = \exp(a)$	$\sigma_z = z\sigma_a$
$z = \ln(a)$	$\sigma_z = \sigma_a/a$	$z = 10^a$	$\sigma_z = \sigma_a/[a\ln(10)]$
$z = a^n$	$\sigma_z = \left\|na^{n-1}\right\|\sigma_a$	$z = \log_{10}(a)$	$\sigma_z = 10^a \ln(10)\sigma_a$
$z = \sin(a)$	$\sigma_z = \|\cos(a)\|\sigma_a$	$z = \cos(a)$	$\sigma_z = \|\sin(a)\|\sigma_a$
$z = a + b$	$\sigma_z = \sqrt{(\sigma_a)^2 + (\sigma_b)^2}$	$z = a - b$	$\sigma_z = \sqrt{(\sigma_a)^2 + (\sigma_b)^2}$
$z = ab$	$\sigma_z = z\sqrt{(\frac{\sigma_a}{a})^2 + (\frac{\sigma_b}{b})^2}$	$z = a/b$	$\sigma_z = z\sqrt{(\frac{\sigma_a}{a})^2 + (\frac{\sigma_b}{b})^2}$

quently, it is only valid when the uncertainties involved are sufficiently small (e.g., less than ~10% to provide a rough estimate [68, 78]).

In the simplest cases, you could develop and run Python functions to propagate errors. Code Listing 10.9 shows two practical examples (i.e., sum and division) based on the rules listed in Table 10.1.

```
1  import numpy as np
2
3  def sum_ab(a, b, sigma_a, sigma_b):
4      z = a + b
5      sigma_z = np.sqrt(sigma_a**2 + sigma_b**2)
6      return z, sigma_z
7
8  def division_ab(a, b, sigma_a, sigma_b):
9      z = a / b
10     sigma_z = z * np.sqrt((sigma_a/a)**2 + (sigma_b/b)**2)
11     return z, sigma_z
```

Listing 10.9 Example application of rules reported in Table 10.1 for sum and division

Also, you could use the symbolic approach to solve Eq. (10.6) or Eq. (10.7). For example, code Listing 10.10 uses SymPy to propagate errors through Eq. (10.6).

```
1  import sympy as sym
2
3  a, b, sigma_a, sigma_b = sym.symbols("a b sigma_a sigma_b")
4
5  def symbolic_error_prop(func, val_a, val_sigma_a, val_b=0,
       val_sigma_b=0):
6
7      z = sym.lambdify([a,b], func, 'numpy')
8      sigma_z = sym.lambdify([a, b, sigma_a, sigma_b], sym.sqrt((sym
         .diff(func, a)**2 * sigma_a**2)+(sym.diff(func, b)**2 *
         sigma_b**2)), 'numpy')
9      val_z = z(a=val_a, b=val_b)
10     val_sigma_z = sigma_z(a=val_a, b=val_b, sigma_a=val_sigma_a,
         sigma_b=val_sigma_b)
11
12     return val_z, val_sigma_z
```

Listing 10.10 Example application of symbolic approach to solving Eq. (10.6)

```
1  my_a = np.array([2, 3, 5, 7, 10])
2  my_sigma_a = np.array([0.2, 0.3, 0.4, 0.7, 0.9])
3  my_b = np.array([2, 3, 6, 4, 8])
4  my_sigma_b = np.array([0.3, 0.3, 0.5, 0.5, 0.5])
5
6  # errors propagated using custom functions
7  my_sum_ab_1, my_sigma_sum_ab_1 = sum_ab(a=my_a, b=my_b, sigma_a=
       my_sigma_a, sigma_b=my_sigma_b)
8  my_division_ab_1, my_sigma_division_ab_1 = division_ab(a=my_a, b=
       my_b, sigma_a=my_sigma_a, sigma_b=my_sigma_b)
9
10 # errors propagated using the symbolic approach
11 my_sum_ab_s, my_sigma_sum_ab_s = symbolic_error_prop(func=a+b,
       val_a=my_a, val_sigma_a=my_sigma_a, val_b=my_b, val_sigma_b=
       my_sigma_b)
```

```
12  my_division_ab_s, my_sigma_division_ab_s = symbolic_error_prop(
        func=a/b, val_a=my_a, val_sigma_a=my_sigma_a, val_b=my_b,
        val_sigma_b=my_sigma_b)
13
14  fig = plt.figure(figsize=(8, 8))
15  ax1 = fig.add_subplot(2, 2, 1)
16  ax1.errorbar(x=my_a, y=my_sum_ab_l, xerr=my_sigma_a, yerr=
        my_sigma_sum_ab_l, linestyle='', marker='o', ecolor='k',
        elinewidth=0.5, capsize=1, label='Errors by custom functions')
17  ax1.set_xlabel('a')
18  ax1.set_ylabel('a + b')
19  ax1.legend()
20  ax2 = fig.add_subplot(2, 2, 2)
21  ax2.errorbar(x=my_a, y=my_sum_ab_s, xerr=my_sigma_a, yerr=
        my_sigma_sum_ab_s, linestyle='', marker='o', ecolor='k',
        elinewidth=0.5, capsize=1, label='Errors by the symbolic
        approach')
22  ax2.set_xlabel('a')
23  ax2.set_ylabel('a + b')
24  ax2.legend()
25  ax3 = fig.add_subplot(2, 2, 3)
26  ax3.errorbar(x=my_a, y=my_division_ab_l, xerr=my_sigma_a, yerr=
        my_sigma_division_ab_l, linestyle='', marker='o', ecolor='k',
        elinewidth=0.5, capsize=1, label='Errors by custom functions')
27  ax3.set_xlabel('a')
28  ax3.set_ylabel('a / b')
29  ax3.legend()
30  ax4 = fig.add_subplot(2,2,4)
31  ax4.errorbar(x=my_a, y=my_division_ab_s, xerr=my_sigma_a, yerr=
        my_sigma_division_ab_s, linestyle='', marker ='o', ecolor='k',
        elinewidth=0.5, capsize=1, label='Errors by the symbolic
        approach')
32  ax4.set_xlabel('a')
33  ax4.set_ylabel('a / b')
34  ax4.legend()
35  fig.tight_layout()
```

Listing 10.11 Error propagation by custom functions reported in code Listing 10.9 and by solving Eq. (10.6) by the symbolic approach (code Listing 10.10)

Finally, code Listing 10.11 and Fig. 10.9 compare the results obtained by propagating errors through custom functions based on the rules listed in Table 10.1 and by the symbolic approach. As expected, the results reported in Fig. 10.9 are identical.

To provide a geological example, consider plotting the ratio Rb/Th versus La for tephras from the recent volcanic activity of the Campi Flegrei Caldera using the linearized approach for error propagation (code Listing 10.12 and Fig. 10.10).

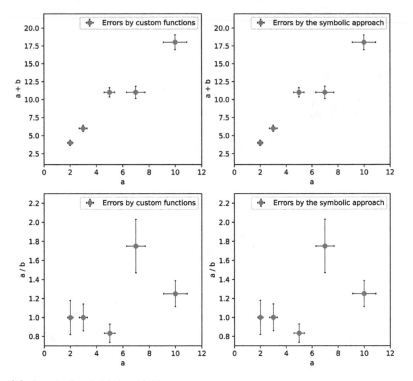

Fig. 10.9 Result of code Listing 10.11

```
1  import pandas as pd
2  import matplotlib.pyplot as plt
3  import sympy as sym
4
5  a, b, sigma_a, sigma_b = sym.symbols("a b sigma_a sigma_b")
6
7  def symbolic_error_prop(func, val_a, val_sigma_a,  val_b=0,
       val_sigma_b=0):
8
9      z = sym.lambdify([a, b], func, 'numpy')
10     sigma_z = sym.lambdify([a, b, sigma_a, sigma_b], sym.sqrt((sym
       .diff(func, a)**2 * sigma_a**2)+(sym.diff(func,b)**2 * sigma_b
       **2)), 'numpy')
11     val_z = z(a=val_a, b=val_b)
12     val_sigma_z = sigma_z(a=val_a, b=val_b, sigma_a=val_sigma_a,
       sigma_b=val_sigma_b)
13
14     return val_z, val_sigma_z
15
16 my_dataset = pd.read_excel('Smith_glass_post_NYT_data.xlsx',
       sheet_name='Supp_traces')
17
18 ratio_y, sigma_ratio_y = symbolic_error_prop(a/b, val_a=my_dataset
       .Rb, val_sigma_a=my_dataset.Rb*0.1, val_b=my_dataset.Th,
       val_sigma_b=my_dataset.Th*0.1)
19
```

```
20  my_dataset['Rb_Th'] = ratio_y
21  my_dataset['Rb_Th_1s'] = sigma_ratio_y
22
23  epochs = ['one','two','three','three-b']
24  colors = ['#afbbb5', '#f10e4a', '#27449c', '#f9a20e']
25
26  fig, ax = plt.subplots()
27  for epoch, color in zip(epochs, colors):
28      my_data = my_dataset[(my_dataset.Epoch == epoch)]
29      ax.errorbar(x=my_data.La, y=my_data.Rb_Th, xerr=my_data.La
            *0.1, yerr=my_data.Rb_Th_1s, linestyle='',  markerfacecolor=
            color, markersize=6, marker='o', markeredgecolor='k', ecolor=
            color, elinewidth=0.5, capsize=0, label="Epoch " + epoch)
30
31  ax.legend(title='CFC Recent Activity')
32  ax.set_ylabel('Rb/Th')
33  ax.set_xlabel('La [ppm]')
```

Listing 10.12 Rb/Th ratio versus La for tephras belonging to recent volcanic activity of Campi Flegrei Caldera. Error propagated using linearized approach

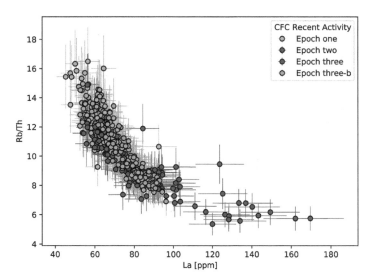

Fig. 10.10 Result of code Listing 10.12

10.4   The Mote Carlo Approach to Error Propagation

Monte Carlo (MC) numerical modeling, named after the casino in the Principality of Monaco, simulates complex probabilistic events using simple random events [64]. MC methods rely on true-random number generators (TRNGs) or pseudo-random number generators (PRNGs) to produce sample distributions simulating a target PDF [64, 69].

What is the difference between TRNGs and PRNGs? TRNGs are devices, generally hardware based, that produce real (i.e., non-deterministic) random numbers [69]. Conversely, PRNGs are deterministic algorithms that generate a "random looking" sequence of numbers [69]. However, given the same starting conditions (i.e., the same seeding), a PRNG will return the same sequence of numbers [69].

In NumPy 1.21, the default PRNG that provides the random sampling of a wide range of distributions (e.g., uniform, normal, etc.) is PCG64.[1] It is a 128-bit implementation of O'Neill's permutation congruential generator [71].

```
1  import numpy as np
2  import matplotlib.pyplot as plt
3
4  def normal_pdf(x, mu, sigma):
5      pdf = 1/(sigma * np.sqrt(2*np.pi)) * np.exp(-(x-mu)**2 / (2*
          sigma**2))
6      return pdf
7
8  def unifrom_pdf(x, a, b):
9      pdf = np.piecewise(x, [(x>=a) & (x<=b), (x<a) & (x>b)], [1/(b-
          a), 0])
10     return pdf
11
12 # Random sampling of a normal distribution
13 my_mu, my_sigma = 0, 0.1 # mean and standard deviation
14 sn = np.random.default_rng().normal(loc=my_mu, scale=my_sigma,
       size=10000)
15 fig = plt.figure()
16 ax1 = fig.add_subplot(2, 1, 1)
17 ax1.hist(sn, density=True, bins='auto', edgecolor='k', color='#
       c7ddf4', label='Random Sampling of the Normal Distribution')
18 my_xn = np.linspace(my_mu - 4 * my_sigma, my_mu + 4 * my_sigma,
       1000)
19 my_yn = normal_pdf(x=my_xn, mu=my_mu, sigma=my_sigma)
20 ax1.plot(my_xn, my_yn, linewidth=2, linestyle='--', color='#ff464a'
       , label='Target Normal Probability Density Function')
21 ax1.set_ylim(0.0, 7.0)
22 ax1.set_xlabel('x')
23 ax1.set_ylabel('Prob. Density')
24 ax1.legend()
25
26 # Random sampling of a uniform distribution
27 my_a, my_b = -1, 1 # lower and upper bound of the uniform
       distribution
28 su = np.random.default_rng().uniform(low=my_a, high=my_b, size
       =10000)
29 ax2 = fig.add_subplot(2, 1, 2)
30 ax2.hist(su, density=True, bins='auto', edgecolor='k', color='#
       c7ddf4', label='Random Sampling of the Uniform Distribution')
31 my_xu = np.linspace(-2, 2, 1000)
32 my_yu = unifrom_pdf(x=my_xu, a=my_a, b=my_b)
33 ax2.plot(my_xu, my_yu, linewidth=2, linestyle='--', color='#ff464a
       ', label='Target Uniform Probability Density Function')
34 ax2.set_ylim(0, 1)
35 ax2.set_xlabel('x')
```

---

[1] https://www.pcg-random.org.

```
36  ax2.set_ylabel('Prob. Density')
37  ax2.legend()
38
39  fig.tight_layout()
```
Listing 10.13 Random sampling of normal and uniform distributions

PCG-64 has a period of 2^{128} and supports advancing an arbitrary number of steps as well as 2^{127} streams.[2]

Code Listing 10.13 provides an example on how to perform a random sampling of specific PDFs by showing how to generate a random sequence of numbers (i.e., a random sample distribution) that simulates normal and uniform PDFs (Fig. 10.11). Code Listing 10.13 uses the *np.random.default_rng()* statement (line 14), which is based on the PCG64 PRNG [71].

Other PRNGs currently available in NumPy are listed in Table 10.2.

Code Listing 10.14 shows how to use a PRNG other than PCG64 to obtain the same normal distribution as in Fig. 10.11 characterized by a mean μ and a standard deviation σ of 0 amd 0.1, respectively. Figure 10.12 shows the results of code Listing 10.14.

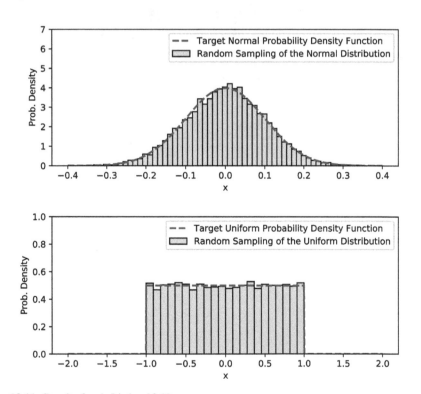

Fig. 10.11 Result of code Listing 10.13

---

[2] https://numpy.org/doc/stable/reference/random/bit_generators/.

Table 10.2 Pseudo random number generators (PRNG) available in NumPy ver. 1.19

PRNG	Reference	Description
PCG64	O'Neill [71]	128-bit implementation of O'Neill's permutation congruential generator
MT19937	Haramoto et al. [67]	Mersenne Twister pseudo-random number generator
Philox	Salmon et al. [75]	64-bit counter-based PRNG using weaker (and faster) versions of cryptographic functions
SFC64	http://pracrand. sourceforge.net	Implementation of Chris Doty-Humphrey's Small Fast Chaotic PRNG

```python
1  import numpy as np
2  import matplotlib.pyplot as plt
3
4  def normal_pdf(x, mu, sigma):
5      pdf = 1/(sigma * np.sqrt(2 * np.pi)) * np.exp( - (x - mu)**2 /
       (2 * sigma**2))
6      return pdf
7
8  fig = plt.figure(figsize=(6,9))
9
10 # Random sampling of a normal distribution
11 my_mu, my_sigma = 0, 0.1 # mean and standard deviation
12
13 bit_generators = [np.random.MT19937(), np.random.Philox(), np.
       random.SFC64()]
14 names = ['Mersenne Twister PRNG (MT19937)', 'Philox (4x64) PRNG (
       Philox)', 'Chris Doty-Humphrey\'s SFC PRNG (SFC64)']
15 indexes = [1,2,3]
16
17 for bit_generator, name, index in zip(bit_generators, names,
       indexes):
18     sn = np.random.Generator(bit_generator).normal(loc = my_mu,
       scale = my_sigma, size = 10000)
19     ax = fig.add_subplot(3, 1, index)
20     ax.hist(sn, density=True, bins='auto', edgecolor='k', color='#
       c7ddf4', label=name)
21     my_xn = np.linspace(my_mu - 4 * my_sigma, my_mu + 4 * my_sigma
       , 1000)
22     my_yn = normal_pdf(x=my_xn, mu=my_mu, sigma=my_sigma)
23     ax.plot(my_xn, my_yn, linewidth=2, linestyle='--', color='#
       ff464a', label ='Target Normal PDF')
24     ax.set_ylim(0.0, 7.0)
25     ax.set_xlim(my_mu - 6 * my_sigma, my_mu + 6 * my_sigma)
26     ax.set_xlabel('x')
27     ax.set_ylabel('Probability Density')
28     ax.legend()
29
30 fig.tight_layout()
```

Listing 10.14 Random sampling of normal distribution created by different PRNGs

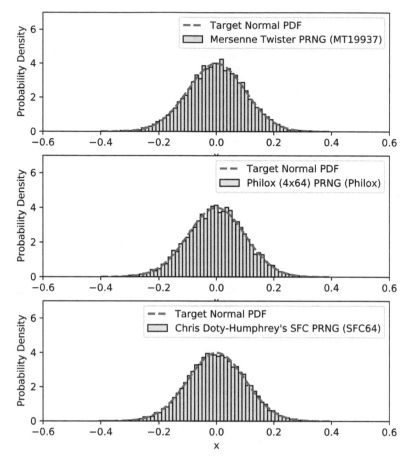

Fig. 10.12 Result of code Listing 10.14

For most basic tasks in geological modeling (e.g., basic error propagation), all PRNGs listed in Table 10.2 work satisfactorily, so I suggest using the default generator for simplicity of use.

In error propagation, the MC approach is a proficient technique to be considered when Eq. (10.6) or its corrected forms (e.g., Eq. 10.7) are inconvenient [76].

```
1   import numpy as np
2   import matplotlib.pyplot as plt
3
4   def gaussian(x, mean, std):
5       return 1/(np.sqrt(2*np.pi*std**2))*np.exp(-0.5*(((x - mean)
        **2)/(std**2)))
6
7   my_a, my_sigma_a = 40, 8
8   my_b, my_sigma_b = 20, 2
9
10  n = 10000
11  a_normal = np.random.default_rng().normal(my_a, my_sigma_a, n)
12  b_normal = np.random.default_rng().normal(my_b, my_sigma_b, n)
```

```
13
14   # Linearized Method
15   my_sum_ab_l, my_sigma_sum_ab_l = sum_ab(a=my_a, b=my_b, sigma_a=
         my_sigma_a, sigma_b=my_sigma_b)
16   my_x = np.linspace(20, 100, 1000)
17   my_sum_ab_PDF = gaussian(x=my_x, mean=my_sum_ab_l, std=
         my_sigma_sum_ab_l)
18
19   # Monte Carlo estimation
20   my_sum_ab_mc =   a_normal + b_normal
21   my_sum_ab_mc_mean = my_sum_ab_mc.mean()
22   my_sigma_sum_ab_mc_std = my_sum_ab_mc.std()
23
24   fig, ax = plt.subplots()
25   ax.hist(my_sum_ab_mc, bins='auto', color='#c7ddf4', edgecolor='k',
         density=True, label= r'a+b sample distribution by MC ($\mu_{a
         +b} = $' + "{:.0f}".format(my_sum_ab_mc_mean) + r'  - 1$\
         sigma_{a+b} = $' + "{:.0f}".format(my_sigma_sum_ab_mc_std) + '
         )')
26   ax.plot(my_x, my_sum_ab_PDF, color='#ff464a', linestyle='--',
         label=r'a+b PDF by linearized error propagation ($\mu_{a+b} =
         $' + "{:.0f}".format(my_sum_ab_l) + r'  - 1$\sigma_{a+b} = $'
         + "{:.0f}".format(my_sigma_sum_ab_l) + ')')
27   ax.set_xlabel('a + b')
28   ax.set_ylabel('Probability Density')
29   ax.legend(title='Error Propagation')
30   ax.set_ylim(0,0.07)
31
32   fig.tight_layout()
```

Listing 10.15 Error propagation by MC

Recall that the application of Eq. (10.6) is based on the following strong assumptions [76]: (a) the errors involved are statistically uncorrelated, (b) the variables involved are independent, and (c) the errors must be sufficiently small relative to the corresponding means. A more difficult problem arises when the derivative elements in Eq. (10.6) or Eq. (10.7) can be solved only with great effort or perhaps not at all [76]. This problem, however, could be attacked by numerical methods such as MC error propagation [76]. To provide a detailed description of the MC method is beyond the scope of this introductory text. Here I limit the discussion to the very simple case of the sum of two variables affected by errors with a normal distribution (code Listing 10.15 and Fig. 10.13). This example highlights the power and simplicity of the MC approach for error propagation. Code Listing 10.15 shows that, after defining a sample distribution for each parameter (lines 11 and 12), the error propagation by MC can be accomplished in one line of code (line 20) without using any additional equation other than the equation of interest, which is the sum in our case.

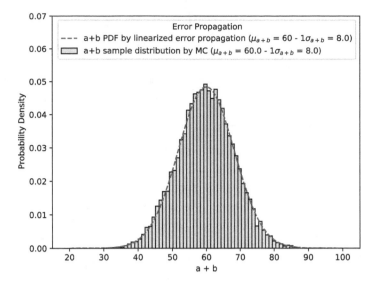

Fig. 10.13   Result of code Listing 10.15

Part V
Robust Statistics and Machine Learning

Chapter 11
Introduction to Robust Statistics

11.1 Classical and Robust Approaches to Statistics

All statistical methods and techniques are based explicitly or implicitly on assumptions [101, 107]. One widely adopted assumption is that the observed (i.e., sampled) data follow a normal (Gaussian) distribution [101, 107]. This assumption is the basis for most classical methods in regression, analysis of variance, and multivariate analysis. However, as is true for many geological cases, although a sample of data may *mostly* follow a normal distribution, some data within the sample may follow a non-normal distribution.

Such atypical data are called outliers. A single outlier can strongly distort statistical methods based on the normal-distribution assumption (e.g., the King-Kong effect in linear regression). Also, if the data are assumed to be normally distributed but their actual distribution diverges from normality, then classical tests may return unreliable results [101, 107].

Definition: "The robust approach to statistical modeling and data analysis aims at deriving methods that produce reliable parameter estimates and associated tests and confidence intervals, not only when the data follow a given distribution exactly, but also when this happens only approximately in the sense just described" [107]. A "robust" model should converge to the results of classical methods when the assumptions behind them (e.g., normal distribution) are satisfied. A complete treatment of robust statistics is beyond the scope of this text, so the interested reader may wish to consult more specialized sources. In the following, I focus on

1. how to check if a sample is normally distributed (i.e., normality tests);
2. robust descriptive statistics;
3. robust linear regression;
4. the application of robust statistics in geochemistry.

11.2   Normality Tests

No standard procedure exists to determine whether a sample follows a normal distribution. However, a reasonable procedure consists of (1) conducting a preliminary qualitative inspection of the histogram plot fit by a normal PDF (see Sect. 9.5), (2) inspecting a quantile-quantile plot, and (3) applying the selected statistical tests of normality [127]. Note that a reasonably large number of observations is needed to detect deviations from normality [101, 107].

Histogram plots and parametric fitting

As reported in Sect. 9.5, plotting the probability density histogram is an easy way to qualitatively determine the shape of a sample distribution. For a normal distribution, we expect the histogram to form a symmetric, bell-shaped curve.

```
1   import pandas as pd
2   import matplotlib.pyplot as plt
3   from scipy.stats import norm
4   import numpy as np
5
6   my_dataset_majors = pd.read_excel('Smith_glass_post_NYT_data.xlsx',
          sheet_name='Supp_majors')
7   my_dataset_traces = pd.read_excel('Smith_glass_post_NYT_data.xlsx',
          sheet_name='Supp_traces')
8
9   fig = plt.figure()
10
11  # MnO
12  MnO = my_dataset_majors.MnO
13
14  ax1 = fig.add_subplot(2, 1, 1)
15  ax1.hist(MnO, bins='auto', density=True, color='#4881e9', edgecolor='k',
          label='MnO', alpha=0.8)
16  a_mean = MnO.mean()
17  std_dev = MnO.std()
18  x = np.linspace(a_mean-4*std_dev, a_mean+4*std_dev,1000)
19  pdf = norm.pdf(x, loc=a_mean, scale=std_dev)
20  ax1.plot(x, pdf, linewidth=1.5, color='#ff464a',label='Normal PDF')
21  ax1.set_xlabel('MnO [wt %]')
22  ax1.set_ylabel('Probability density')
23  ax1.legend()
24
25  #Pb
26  Pb = my_dataset_traces.Pb
27  Pb = Pb.dropna(how='any')
28  ax2 = fig.add_subplot(2, 1, 2)
29  ax2.hist(Pb, bins='auto', density=True, color='#4881e9', edgecolor='k',
          label='Pb', alpha=0.8)
30  a_mean = Pb.mean()
31  std_dev = Pb.std()
32  x = np.linspace(a_mean-4*std_dev, a_mean+4*std_dev,1000)
33  pdf = norm.pdf(x, loc=a_mean, scale=std_dev)
34  ax2.plot(x, pdf, linewidth=1.5, color='#ff464a', label='Normal PDF')
35  ax2.set_xlabel('Pb [ppm]')
36  ax2.set_ylabel('Probability density')
37  ax2.legend()
38
39  fig.align_ylabels()
40  fig.tight_layout()
```

Listing 11.1 Histogram of distribution with a parametric fit to assess the normality of the sample distribution

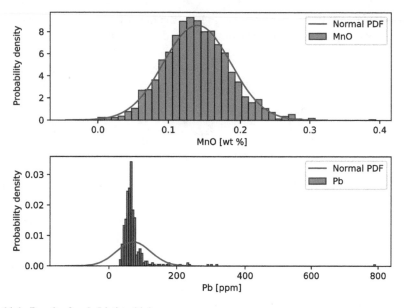

Fig. 11.1 Result of code Listing 11.1

Fitting parametrically to a normal PDF then allows us to better evaluate the similarities and differences between the sample studied and a normal distribution characterized by the same mean and standard deviation.

For example, consider the distributions of the MnO and Pb data sets reported by Smith et al. [16], which clearly depart from a normal distribution for Pb: note the tail extending to the right, which gives a positive skewness, and the strong outlier near 790 ppm (code Listing 11.1 and Fig. 11.1). Conversely, the MnO probability density histogram is nearly symmetric with no outliers except for a single data point near 0.39 wt %.

The parametric fitting of the two distributions with a Gaussian PDF (code Listing 11.1 and Fig. 11.1) confirms a strong departure from normality for Pb and a near-normal distribution for MnO.

Given that this is a qualitative analysis, the plot of the density histogram and the parametric fitting to a normal distribution can only detect significant departures from a normal distribution. Consequently, although we can state with certainty that the Pb sample does not follow a normal distribution, we cannot do the same for the MnO distribution [127].

Quantile-Quantile plots

The next step in the investigation of normality of a sample distribution is a quantile-quantile (Q-Q) plot [115]. The Q-Q plot is a graphical representation used to determine whether two data sets come from populations characterized by the same distribution. When used to test for the normality of a sample distribution, one of the two data sets serves as the investigated sample, and the other derives from a normal PDF.

In detail, we develop a binary diagram where the quantiles of the investigated data set are plotted against the quantiles of a normal distribution.

```
1  import statsmodels.api as sm
2
3  fig = plt.figure()
4
5  ax1 = fig.add_subplot(1, 2, 1)
6  sm.qqplot(data=MnO, fit = True, line="45", ax=ax1, markerfacecolor='#4881
       e9', markeredgewidth='0.5', markeredgecolor='k', label='MnO')
7  ax1.set_aspect('equal', 'box')
8  ax1.legend(loc='lower right')
9
10 ax2 = fig.add_subplot(1, 2, 2)
11 sm.qqplot(data=Pb, fit = True, line="45", ax=ax2, markerfacecolor='#4881
       e9', markeredgewidth='0.5', markeredgecolor='k', label='Pb')
12 ax2.set_aspect('equal', 'box')
13 ax2.legend(loc='lower right')
14
15 fig.tight_layout()
```

Listing 11.2 Q-Q diagrams for MnO and Pb

If the investigated data set comes from a population with a normal distribution, the standardized quantiles (i.e., derived after subtracting the mean and dividing by the standard deviation) should fall approximately along a 1 : 1 reference line.

The greater the departure from this reference line, the greater is the evidence that the investigated data set does not come from a normal population. As an example, Fig. 11.2 (code Listing 11.2) shows Q-Q plots for the MnO and Pb samples. As expected, the Q-Q plot for Pb departs strongly from the reference line, demonstrating further the non-normality of the sample. The Q-Q plot for MnO, however, shows that the sample quantiles are mostly consistent with the theoretical quantiles. However, at least one observation (the outliers at 0.39 wt % on the extreme right side of Fig. 11.1) in the Q-Q plot departs from linearity. Can we assume that MnO follows a normal distribution? Answering this question requires further statistical tests.

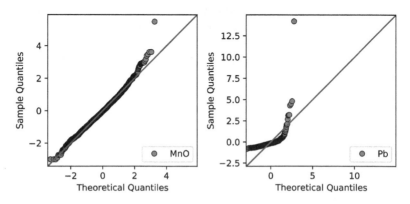

Fig. 11.2 Result of code Listing 11.2

Statistical tests

Typically, a statistical test for normality initially assumes that the sample derives from a normal (Gaussian) population [127]. This initial assumption is the so-called null hypothesis H_0. Tests then elaborate data and return one or more statistical parameters and one or more threshold values to determine whether we can accept H_0 [127].

The Shapiro-Wilk (S-W) test is a statistical procedure for testing a sample data set for normality [124]. Specifically, the S-W test relies on the W parameter, which is determined by dividing the square of an appropriate linear combination of the sample order statistics by the usual symmetric estimate of variance [124]. The maximum value of W is unity, corresponding to a normal distribution. Thus, the closer W is to unity, the closer the sample approaches a normal distribution. A small value for W indicates that the sample is not normally distributed. In practice, you can reject the null hypothesis if W is less than a certain threshold.

The D'Agostino-Pearson (DA-P) test evaluates two descriptive statistics, the skewness and kurtosis, to test for normality [91, 92]. In detail, this test estimates the p-value, combining the two metrics to quantify the departure from a Gaussian distribution [91, 92]. As with the S-W test, you can reject the null hypothesis that your population is normally distributed if the p-value is less than a certain threshold.

The Anderson-Darling (A-D) test is a modification of the Kolmogorov-Smirnov (K-S) test [126]. Rather than returning a single p-value as in the case of the DA-P test, the A-D test returns statistics (i.e., a series of computed values and a list of critical values). If the returned statistic exceeds the reference critical value, then, for the given significance level, the null hypothesis that the data come from the chosen distribution (the normal distribution in our case) can be rejected [126].

Code Listing 11.3 shows how to implement S-W, DA-P, and A-D tests in Python for a geological data set.

```
1   def returns_normal_tests(my_data):
2
3       from scipy.stats import shapiro, anderson, normaltest
4
5       print('------------------------------------------------')
6       print('')
7       stat, p = shapiro(my_data)
8       alpha = 0.05
9       if p > alpha:
10          print('Shapiro test fails to reject H0: looks normal ')
11      else:
12          print('Shapiro test rejects H0: not normal ')
13      print('')
14      stat, p = normaltest(my_data)
15      alpha = 0.05
16      if p > alpha:
17          print("D'Agostino and Pearson's test fails to reject H0: looks
            normal")
18      else:
19          print("D'Agostino and Pearson's test rejects H0: not normal")
20      print('')
21      result = anderson(my_data)
22      print('Anderson-Darling test:')
23      for sl, cv in zip(result.significance_level, result.critical_values):
```

```
24            if result.statistic < cv:
25                print('%.3f: fails to reject H0: Sample looks normal' % (s1))
26            else:
27                print('%.3f: rejects H0: Sample does not look normal' % (s1))
28        print('-----------------------------------------------')
29        print('')
30
31 # Original MnO sample
32 print('Original MnO sample')
33 returns_normal_tests(MnO)
34
35 # Removing the outliers above 0.27 wt %
36 print('MnO sample without observations above 0.27 wt %')
37 MnO_no_outliers = MnO[MnO < 0.27]
38 returns_normal_tests(MnO_no_outliers)
39
40 ''' Results:
41 Original MnO sample
42 ------------------------------------------------
43
44 Shapiro test rejects H0: not normal
45
46 D'Agostino and Pearson's test rejects H0: not normal
47
48 Anderson-Darling test:
49 15.000: rejects H0: Sample does not look normal
50 10.000: rejects H0: Sample does not look normal
51 5.000: rejects H0: Sample does not look normal
52 2.500: rejects H0: Sample does not look normal
53 1.000: rejects H0: Sample does not look normal
54 ------------------------------------------------
55
56 MnO sample without observations above 0.27 wt %
57 ------------------------------------------------
58
59 Shapiro test fails to reject H0: looks normal
60
61 D'Agostino and Pearson's test fails to reject H0: looks normal
62
63 Anderson-Darling test:
64 15.000: fails to reject H0: Sample looks normal
65 10.000: fails to reject H0: Sample looks normal
66 5.000: fails to reject H0: Sample looks normal
67 2.500: fails to reject H0: Sample looks normal
68 1.000: fails to reject H0: Sample looks normal
69 -------------------------------------------------
70 '''
```

Listing 11.3 Performing statistical tests of normality for the MnO sample

11.3 Robust Estimators for Location and Scale

Chapter 5 reviews the classical estimators of location and scale (or spread) for a sample distribution, which are the building blocks of descriptive statistics. Examples include the sample mean and standard deviation as estimators for the location and scale, respectively. However, outliers may cause these estimators to fail. In such cases, robust estimators are a better choice [101, 107]. The following provides a brief introduction to robust estimators for the location and scale of univariate sample distributions and their implementation in Python. The interested reader may consult more specialized books for a more thorough treatment of the topic [101, 107].

Robust and weak estimators for location

Of the classical estimators for location, the arithmetic mean is the most used and the most widely recognized (cf. Chap. 5). However, the arithmetic mean is strongly affected by outliers [101, 107]. For example, considering the Pb distribution in the data set reported by Smith et al. [16], we find a positive tail and a strong outlier at 790 ppm (Fig. 11.3). The arithmetic mean for Pb is 81 ppm; this is greater than most observations, which range from 50 to 80 ppm (Fig. 11.3). This result is due to the strong influence on the arithmetic mean of positive outliers. Consequently, the arithmetic mean is considered a weak estimator of location in the presence of outliers.

In contrast, the median is 67 ppm (Fig. 11.3), which is centered within the interval containing most observations (i.e., 50–80 ppm) and corresponds to the modal bin in Fig. 11.3. This is because the median is less affected by outliers than the arithmetic mean, making the median a robust estimator for location in the presence of outliers.

Another approach to obtain a robust estimate for the location of a sample distribution is through the trimmed mean [101, 107], which consists in defining a criterion to discard a fraction of the largest and smallest values, as follows: let $\alpha \in [0, 1/2]$ and $m = [n\alpha]$, where $[\cdot]$ returns the integer part and n is the total number of observations. We define the α-trimmed mean as [101, 107]

$$\mu_\alpha = \bar{z}_\alpha = \frac{1}{n - 2m} \sum_{i=m+1}^{n-m} z_{(i)}, \tag{11.1}$$

where $z_{(i)}$ denotes ordered observations. The limiting cases $\alpha = 0$ and $\alpha \to 0.5$ correspond to the sample mean and median, respectively.

The α-Winsorized mean μ_{Wins} is similar to the α-trimmed mean but, instead of deleting extreme values as in the trimmed mean, it shifts them toward the bulk of the data [Eq. (11.2)]:

```
1   import pandas as pd
2   import numpy as np
3   from scipy.stats.mstats import winsorize
4   from scipy.stats import trim_mean
5   import matplotlib.pyplot as plt
6
7   my_dataset = pd.read_excel('Smith_glass_post_NYT_data.xlsx', sheet_name
        =1)
8
9   el = 'Pb'
10  my_sub_dataset = my_dataset[my_dataset.Epoch == 'three-b']
11  my_sub_dataset = my_sub_dataset.dropna(subset=[el])
12
13  fig, ax = plt.subplots()
14  a_mean = my_sub_dataset[el].mean()
15  median = my_sub_dataset[el].median()
16  trimmed_mean = trim_mean(my_sub_dataset[el], proportiontocut=0.1)
```

```
17  winsorized_mean = np.mean(winsorize(my_sub_dataset[el], limits=0.1))
18
19  delta = 100 * (a_mean-median) / median
20
21  bins = np.arange(50,240,5)
22  ax.hist(my_sub_dataset[el], density=True, edgecolor='k', color='#4881e9'
        , bins=bins, label = 'Lead (Pb), Epoch Three')
23  ax.axvline(a_mean, color='#ff464a', linewidth=2, label='Arithmetic Mean:
        {:.0f} [ppm]'.format(a_mean))
24  ax.axvline(median, color='#ebb60d', linewidth=2, label='Median: {:.0f} [
        ppm]'.format(median))
25  ax.axvline(trimmed_mean, color='#8f10b3', linewidth=2, label=r'Trimmed
        Mean ($\alpha = 0.1$):' + '{:.0f} [ppm]'.format(trimmed_mean))
26  ax.axvline(winsorized_mean, color='#07851e', linewidth=2, label=r'
        Winsored Mean ($\alpha = 0.1$):' + '{:.0f} [ppm]'.format(
        winsorized_mean))
27
28  ax.set_xlabel(el + " [ppm]")
29  ax.set_ylabel('probability density')
30  ax.legend()
31  ax.annotate('Large oulier at about 800 ppm', (240, 0.02), (220, 0.02), ha
        ="right", va="center", size=9, arrowprops=dict(arrowstyle='fancy'))
32  ax.annotate('Deviation of the arithmetic\nmean from the median: {:.1f} %'
        .format(delta), (a_mean + 3, 0.03), (a_mean + 25, 0.03), ha="left",
        va="center", size=9, arrowprops=dict(arrowstyle='fancy'))
```

Listing 11.4 Weak and robust estimates for location

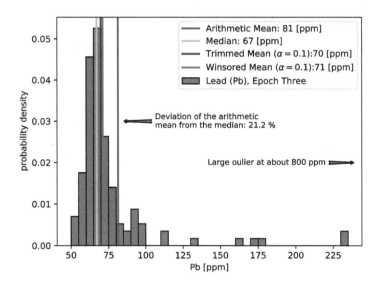

Fig. 11.3 Result of code Listing 11.4

$$\mu_{\text{Wins}} = \frac{1}{n} \left(m z_{(m)} + m z_{(n-m+1)} + \sum_{i=m+1}^{n-m} z_{(i)} \right), \qquad (11.2)$$

where m and $z_{(i)}$ are defined as for the trimmed mean [Eq. (11.1)]. In Python, the trimmed and Winsorized means can be easily estimated by using the *trim_mean()* and *winsorize()* methods in scipy.stats and scipy.stats.mstats, respectively (code Listing 11.4 and Fig. 11.3).

Robust and weak estimators for the scale

Chapter 5 also reviewed the main estimators for the scale of a distribution. One of the weaker scale estimators is the range (Fig. 11.4). In addition, the standard deviation is strongly affected by the presence of outliers (Fig. 11.4). Another scale estimator discussed in Chap. 5 is the Inter-Quartile Range (IQR; Fig. 11.4). Here, I introduce an additional robust scale estimator called the "median absolute deviation" (MAD) about the median, which is defined as

$$\text{MAD}(\mathbf{z}) = \text{MAD}(z_1, z_2, \ldots, z_n) = \text{Me}\left\{|\mathbf{z} - \text{Me}(\mathbf{z})|\right\}. \qquad (11.3)$$

```
1  import pandas as pd
2  import numpy as np
3  from scipy import stats
4  import matplotlib.pyplot as plt
5
6  my_dataset = pd.read_excel('Smith_glass_post_NYT_data.xlsx', sheet_name
       =1)
7  el = 'Pb'
8  my_sub_dataset = my_dataset[my_dataset.Epoch == 'three-b']
9  my_sub_dataset = my_sub_dataset.dropna(subset=[el])
10  a_mean = my_sub_dataset[el].mean()
11  median = my_sub_dataset[el].median()
12  range_values = [my_sub_dataset[el].min(), my_sub_dataset[el].max()]
13  std_dev_values = [a_mean - my_sub_dataset[el].std(), a_mean +
       my_sub_dataset[el].std()]
14  IQR_values = [np.percentile(my_sub_dataset[el], 25, interpolation = '
       midpoint'), np.percentile(my_sub_dataset[el], 75, interpolation = '
       midpoint')]
15  MADn_values = [median - stats.median_abs_deviation(my_sub_dataset[el],
       scale='normal'), median + stats.median_abs_deviation(my_sub_dataset[
       el], scale='normal')]
16
17  scales_values = [range_values, std_dev_values, IQR_values, MADn_values]
18  scale_labels = ['Range', 'Standard Deviation', 'Inter Quartile Range', '
       Median Absolute Deviation']
19  locations = [a_mean, a_mean, median, median]
20  location_labels = ['Arithmetic Mean', 'Arithmetic Mean', 'Median', '
       Median']
21  binnings = ['auto', np.arange(0,300,5),np.arange(50,150,5),np.arange
       (50,150,5)]
22  indexes = [1,2,3,4]
23
24  fig = plt.figure(figsize=(8,6))
```

```
25  for  scale_values,  location,  scale_label,  location_label,  bins,  index  in
         zip(scales_values,  locations,  scale_labels,  location_labels,
         binnings,  indexes):
26      ax  =  fig.add_subplot(2,  2,  index)
27      ax.hist(my_sub_dataset[el],  density=True,  edgecolor='k',  color='#4881
         e9',  bins=bins)
28      ax.axvline(location,  color='#ff464a',  linewidth=1,  label=
         location_label)
29      ax.axvline(scale_values[0],  color='#ebb60d')
30      ax.axvline(scale_values[1],  color='#ebb60d')
31      ax.axvspan(scale_values[0],  scale_values[1],  alpha=0.1,  color='orange
         ',  label=scale_label)
32      ax.set_xlabel(el  +  "  [ppm]")
33      ax.set_ylabel('probability  density')
34      ax.set_ylim(0,  0.1)
35      ax.legend(loc  =  'upper  right')
36  fig.tight_layout()
```

Listing 11.5 Weak and robust estimates for scale

The MAD uses the sample median twice, first to estimate the location of the data set [i.e., $Me(\mathbf{z})$], and then to compute the sample median of the absolute residuals from the estimated location [i.e., $\{|\mathbf{z} - Me(\mathbf{z})|\}$]. To make the MAD comparable to σ, the normalized MAD (MAD_n) is defined as

$$MAD_n(\mathbf{z}) = \frac{MAD(\mathbf{z})}{0.6745}. \tag{11.4}$$

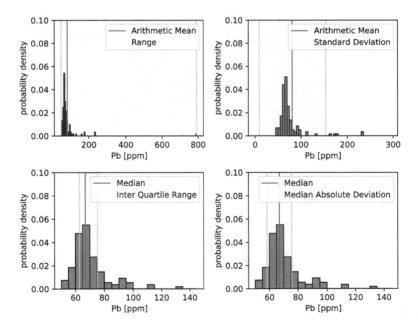

Fig. 11.4 Result of code Listing 11.5

The rationale behind this choice is that the number 0.6745 is the MAD of a standard normal random variable, so a variable $N(\mu, \sigma)$ has $MAD_n = \sigma$. In Python, the MAD can be easily computed by using the *scipy.stats.median_abs_deviation()* function. To calculate the MAD_n as defined by Eq. (11.4), we need to explicitly set the *scale* parameter to 'normal' when calling the *median_abs_deviation()* function.

M estimators of location and scale

The jointly robust estimate of location and scale proposed by Huber (1966) (i.e., "Huber's proposal 2") consists of the solution of a location–dispersion model with two unknown parameters $\hat{\mu}$ and $\hat{\sigma}$:

```
1   import pandas as pd
2   import numpy as np
3   import statsmodels.api as st
4   import matplotlib.pyplot as plt
5
6   my_dataset = pd.read_excel('Smith_glass_post_NYT_data.xlsx', sheet_name
        =1)
7   el = 'Pb'
8
9   my_sub_dataset = my_dataset[my_dataset.Epoch == 'three-b']
10  my_sub_dataset = my_sub_dataset.dropna(subset=[el])
11
12  norms = [st.robust.norms.HuberT(t=1.345), st.robust.norms.Hampel(a=2.0, b
        =4.0, c=8.0)]
13  loc_labels = [r"Huber's T function", r"Hampel function"]
14  indexes = [1,2]
15
16  fig = plt.figure(figsize=(6,6))
17
18  for norm, loc_label, index in zip(norms, loc_labels, indexes):
19
20      huber_proposal_2 = st.robust.Huber(c= 1.5, norm = norm)
21      h_loc, h_scale = huber_proposal_2(my_sub_dataset[el])
22      ax = fig.add_subplot(2, 1, index)
23      bins = np.arange(50,250,5)
24      ax.hist(my_sub_dataset[el], density = True, edgecolor='k', color='
            #4881e9', bins=bins)
25      ax.axvline(h_loc, color = '#ff464a', linewidth = 2, label= loc_label
            + " as $\psi$: location at {:.1f} [ppm]".format(h_loc))
26      ax.axvline(h_loc + h_scale, color = '#ebb60d')
27      ax.axvline(h_loc - h_scale, color = '#ebb60d')
28      ax.axvspan(h_loc + h_scale, h_loc - h_scale, alpha=0.1, color='orange
            ', label="Huber's estimation for the scale: {:.1f} [ppm]".format(
            h_scale))
29      ax.set_xlabel(el + " [ppm]")
30      ax.set_ylabel('probability density')
31      ax.set_ylim(0, 0.1)
32      ax.legend(loc = 'upper right')
33      ax.annotate('Large oulier at about 800 ppm', (253, 0.04), (230,0.04),
            ha='right', va='center', size=9, arrowprops=dict(arrowstyle='fancy'
            ))
34  fig.tight_layout()
```

Listing 11.6 M estimators for location and scale: "Huber's proposal 2"

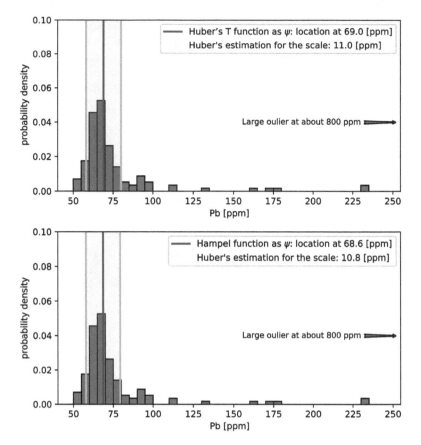

Fig. 11.5   Result of code Listing 11.6

$$\sum_{i=1}^{n} \psi \left(\frac{x_1 - \hat{\mu}}{\hat{\sigma}} \right) = 0,$$

$$\sum_{i=1}^{n} \psi^2 \left(\frac{x_1 - \hat{\mu}}{\hat{\sigma}} \right) = (n-1)\beta, \qquad (11.5)$$

where $\hat{\mu}$ and $\hat{\sigma}$ are the maximum likelihood estimators of μ and σ, respectively. In Python, Huber's proposal 2 is implemented by the *statsmodels.robust.scale.Huber()* function. By default, it uses the Huber's T as ψ, but other ψ can be selected (e.g., Hampel 17A, Ramsay's Ea, etc.) (see code Listing 11.6 and Fig. 11.5).

11.4   Robust Statistics in Geochemistry

The present section reviews the main conclusions reported by [73] on the use of robust statistics in geochemistry. For example, Reimann and Filzmoser [73] claim that most of the variables in large data sets from regional geochemical and environmental surveys show neither normal nor log-normal distribution.

Even after a transformation devoted to producing a normal-distribution data set, many of these data sets do not approach a Gaussian distribution [73]. Typically, the distributions investigated by Reimann and Filzmoser [73] are skewed and contain outliers. Reimann and Filzmoser [73] concluded that, when dealing with regional geochemical or environmental data, normal or log-normal distributions are an exception and not the rule. The conclusions reported by Reimann and Filzmoser [73] have significant consequences for the further statistical treatment of geochemical and environmental data, mostly requiring a robust approach.

Why are geochemical and environmental data not normally distributed? Reimann and Filzmoser [73] argue that geochemical and environmental data have a spatial dependence, and spatially dependent data typically are not normally distributed. Also, trace-element data approaching the detection limit are often truncated, which means that a significant number of observations are not characterized by a true measured value [73]. Furthermore, the precision of the analytical determinations deteriorates as the element concentration decreases, so values are less precise when approaching detection limits [73]. Finally, these data sets often contain outliers, possibly due to analytical issues or due to a population other than the main population of the data [73].

Table 11.1 Application of robust statistics in geochemistry. Developed from Table 3 of Reimann and Filzmoser [73]

Location	Recommendation	Here
Arithmetic mean	Should only be used in special cases	Yes
Geometric mean	Can be used, but may be problematic in some cases	Yes
Median	Should be the first choice as location estimator	Yes
Hampel or Huber means	Can be used	Yes
Dispersion	**Recommendation**	**Here**
Standard deviation	Should not be used if data outliers exist	Yes
Mad (medmed)	Can be used	Yes
Hinge spread	Can be used	No
Robust spread	Can be used	Yes
Tests for means and variances	**Recommendation**	**Here**
t-test	Should not be used	No
F-test	Should not be used	No
Notches in boxplot	Can be used, very easy and fast	Yes
Non-parametric tests	Can be used	Yes
Robust tests	Can be used	No

(continued)

Table 11.1 (continued)

Multivariate methods	Recommendation	Here
Correlation analysis	Should not be used with the original (untransformed) data	Yes
Regression analysis	Should not be used with the original (untransformed) data	Yes
Robust regression analysis	Can be used, preferably on log-transformed data	No
Non-parametric regression	Can be used, preferably on log-transformed data	Yes
PCA	Very sensible to outlying observations, Should not be used	No
Robust PCA	Can be used, preferably with log-transformed data	No

Table 11.1 is a modification of Table 3 from Reimann and Filzmoser [73] and lists the frequently used statistical parameters, tests, and multivariate methods and their suitability for regional geochemical and environmental data that have neither a normal or log-normal distribution.

Chapter 12
Machine Learning

12.1 Introduction to Machine Learning in Geology

Machine learning (ML) is a sub-field of Artificial Intelligence (AI) and concerns the use of algorithms and methods to detect patterns in large data sets and the use these patterns to predict future trends, to classify, or to make other types of strategic decisions [110].

The field of ML has grown significantly over the past two decades, evolving from a "niche approach" to a robust technology with broad scientific and commercial use [103]. For example, ML is now used in several fields such as speech recognition, computer vision, robot control, and natural language processing [103]. In principle, any complex problem described by a sufficiently large number of input samples and features may be treated by ML [103]. Over the last decade, numerous researchers have started investigating the application of ML methods in the Earth Sciences [84, 90, 96, 100, 109, 117–120, 129]. This section introduces the basics of ML in Python and highlights a case study in the field of Earth Sciences.

A common characteristic of ML applications is that they are not developed to process a conceptual model defined a priori but instead attempt to uncover the complexities of large data sets through a so-called learning process [86, 123]. The goal of the process is to convert experience into "expertise" or "knowledge" [123]. Note that this is analogous to how humans learn from past experiences.

For example, children begin learning the alphabet by observing the world around them where they find sounds, written letters, words, or phrases. Then, at school, they learn the significance of the alphabet and how to combine the different letters. Similarly, the experience for a ML algorithm is the training data and the output is the learned expertise, such as a model that can perform a specific task [123].

Broadly speaking, the learning process in ML can be divided into two main fields: (a) unsupervised learning and (b) supervised learning. In unsupervised learning, the training data set consists of several input vectors or arrays, with no corresponding

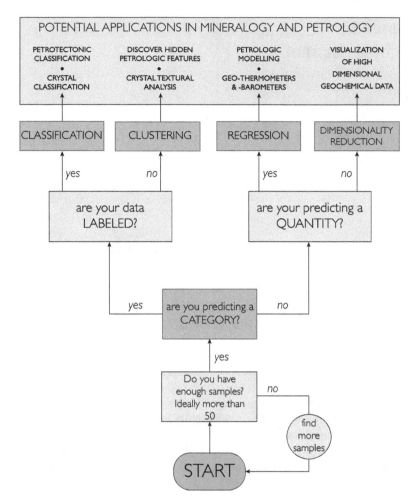

Fig. 12.1 Workflow for the application of ML techniques in petrology and mineralogy [modified from [119]]

target values. Conversely, in supervised learning, the training data set is labeled, meaning that the algorithm learns through examples [86].

Figure 12.1 shows a flowchart, modified from Petrelli and Perugini [119], depicting the main areas of ML (classification, clustering, regression, and dimensionality reduction) and their possible use to solve typical mineralogical and petrological problems. As shown in Fig. 12.1, the use of ML requires the availability of a significant number of data (ideally more than 50[1]). The main aim of Fig. 12.1 is to determine which ML field is best suited to approach the problem (classification, clustering, regression, or dimensionality reduction). The procedure entails a range of choices about the nature of the question under investigation.

---

[1] https://scikit-learn.org/stable/tutorial/machine_learning_map/index.html.

If the problem involves categories, the first step is to select between labeled and unlabeled data. If the learning data set is labeled, the training process is supervised and will involve a "classification" problem [104]. An example of a classification problem in petrology is petro-tectonic identification using geochemical data [119]. If the training data set is unlabeled, the problem is about "clustering" [102]. The field of clustering has been investigated in petrology since the 1980s [105]. For example, Le Maitre [105] discusses the basics of clustering in petrology. If the problem does not include a categorization, the next step is to establish whether a quantity must be predicted. If so, we are in the field of "regression" [125]. An example of an application in petrology of ML regression is provided by Petrelli et al. [118]. Finally, if we are not predicting a quantity, we are in the field of "dimensionality reduction" [106]. Dimensionality reduction is particularly useful, for example, in the context of visualization of high-dimensional geological data.

12.2 Machine Learning in Python

To introduce the application of ML techniques to Earth Sciences, I will use Scikit-learn.[2] Scikit-learn is a Python library that contains a wide range of state-of-the-art ML algorithms [116]. This package focuses on bringing ML to non-specialists via a general-purpose high-level language such as Python [116].

Scikit-learn is a robust framework for solving Earth Sciences problems in the fields of clustering, regression, dimensionality reduction, and classification (Fig. 12.1). Other examples of Python libraries for the development of ML applications are TensorFlow,[3] Keras,[4] and PyTorch.[5]

12.3 A Case Study of Machine Learning in Geology

Pyroxene thermobarometry
Determining pre-eruptive temperatures and storage depths in volcanic plumbing systems is a fundamental issue in petrology and volcanology [93, 121, 122]. To date, the development of geo-thermometers and barometers has been based on the thermodynamic characterization of the magmatic system, which provides a robust framework that is widely applied to estimate pre-eruptive magma temperature and storage depths [108, 112–114, 121, 122]. As reported by Petrelli et al. [118], the conventional calibration procedure for CPX thermometers and barometers consists of five main steps:

---

[2] https://scikit-learn.org.

[3] https://www.tensorflow.org.

[4] https://keras.io.

[5] https://pytorch.org.

1. recognize chemical equilibria associated with large variations of entropy and volume [121];
2. retrieve a statistically robust experimental data set with known T and P (e.g., the LEPR data set [98]);
3. determine the CPX components from chemical analyses;
4. define a regression procedure;
5. validate the model [121].

In 2020, Petrelli et al. [118] proposed a new ML method to retrieve magma temperature and storage depths on the basis of melt-clinopyroxenes and clinopyroxene-only chemistry. The ML approach proposed by Petrelli et al. [118] starts from the same basis as the classical approach but is not based on a model defined a priori, thereby allowing the algorithm to retrieve the elements that are involved in variations of entropy and volume. But what is the main difference between classical approaches and ML approaches? In a few words, classical approaches are based on a simplified thermodynamic framework that provides equations with which to fit the experimental data (typically using linear regression). Conversely, ML methods are based on the statistical relationships that relate variations in the chemistry of CPXs (or CPX-melt couples) to the target variables (i.e., P and T), without necessarily providing a thermodynamic framework. In agreement with the workflow reported in Fig. 12.1, the investigations of Petrelli et al. [118] fall into the ML field of regression.

Experimental data set for training

The experimental data set used by Petrelli et al. [118] to train the model consisted of 1403 experimentally produced clinopyroxenes in equilibrium with a wide range of silicate melt compositions at pressures and temperatures in the range 0.001–40 kbar and 952–1883 K. As input parameters, Petrelli et al. [118] used the major element compositions of melt (SiO_2, TiO_2, Al_2O_3, FeO_t, MnO, MgO, CaO, Na_2O, K_2O, Cr_2O_3, P_2O_5, H_2O) and clinopyroxene (SiO_2, TiO_2, Al_2O_3, FeO_t, MnO, MgO, CaO, Na_2O, K_2O, Cr_2O_3) phases. We now import and visualize the data set shared by Petrelli et al. [118] by using code Listing 12.1 and Figs. 12.2 and 12.3.

```
1  import numpy as np
2  import pandas as pd
3  import matplotlib.pyplot as plt
4  import seaborn as sns
5  from sklearn.preprocessing import StandardScaler
6  from sklearn.ensemble import ExtraTreesRegressor
7  from sklearn.metrics import mean_squared_error
8  from sklearn.metrics import r2_score
9
10 # Import The Training Data Set
11 my_training_dataset = pd.read_excel('
      GlobalDataset_Final_rev9_TrainValidation.xlsx',
         usecols = "A:M,O:X,Z:AA", skiprows=1, engine='
      openpyxl')
```

```
12 my_training_dataset.columns = [c.replace('.1', 'cpx
       ') for c in my_training_dataset.columns]
13 my_training_dataset = my_training_dataset.fillna(0)
14
15 train_labels = np.array([my_training_dataset.
       Sample_ID]).T
16 X0_train = my_training_dataset.iloc[:, 1:23]
17 Y_train = np.array([my_training_dataset.T_K]).T
18
19 fig = plt.figure(figsize=(8,8))
20 x_labels_melt = [r'SiO$_2$', r'TiO$_2$', r'
       Al$_2$O$_3$', r'FeO$_t$', r'MnO', r'MgO', r'CaO
       ', r'Na$_2O$', r'K$_2$O', r'Cr$_2$O$_3$', r'
       P$_2$O$_5$', r'H$_2$O']
21 for i in range(0,12):
22     ax1 = fig.add_subplot(4, 3, i+1)
23     sns.kdeplot(X0_train.iloc[:, i],fill=True,
       color='k', facecolor='#c7ddf4', ax = ax1)
24     ax1.set_xlabel(x_labels_melt[i] + ' [wt. %] the
       melt')
25 fig.align_ylabels()
26 fig.tight_layout()
27
28 fig1 = plt.figure(figsize=(6,8))
29 x_labels_cpx = [r'SiO$_2$', r'TiO$_2$', r'
       Al$_2$O$_3$', r'FeO$_t$', r'MnO', r'MgO', r'CaO
       ', r'Na$_2O$', r'K$_2$O', r'Cr$_2$O$_3$']
30 for i in range(0,10):
31     ax2 = fig1.add_subplot(5, 2, i+1)
32     sns.kdeplot(X0_train.iloc[:, i+12],fill=True,
       color='k', facecolor='#c7ddf4', ax = ax2)
33     ax2.set_xlabel(x_labels_cpx[i] + ' [wt. %] in
       cpx')
34 fig1.align_ylabels()
35 fig1.tight_layout()
```

Listing 12.1 Importing and visualizing the training data set from Petrelli et al. [118]

Standardization

A standardized data set is a common requirement for many ML estimators.

For instance, many ML algorithms assume that all features are centered on zero and that their variance is of the same order. If a feature has a variance that is orders of magnitude greater than the others, it might play a dominant role and prevent the algorithm from correctly learning other features.

The easiest way to normalize a data set is to subtract the mean and scale to unit variance (Eq. 12.1):

$$\tilde{x}_e^i = \frac{x_e^i - \mu^e}{\sigma_p^e}, \tag{12.1}$$

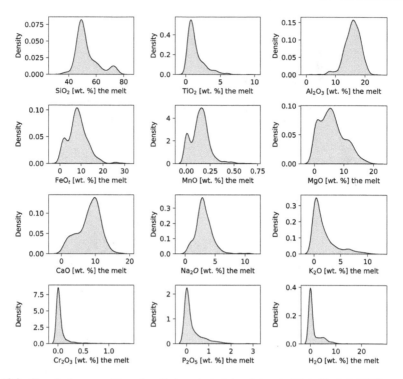

Fig. 12.2 Chemical composition of melt phase in training data set of Petrelli et al. [118]

where $\tilde{x}_e^i$ and x_e^i are the transformed and original components, respectively, belonging to the sample distribution of the chemical analysis of element e (i.e., SiO_2, TiO_2, etc.), which is characterized by a mean μ^e and a standard deviation σ_p^e.

Scikit-learn implements Eq. (12.1) in the *sklearn.preprocessing.StandardScaler()* class, which is a set of methods (i.e., functions) to scale both the training data set and unknown samples.

In addition, scikit-learn implements additional scalers and transformers. In scikit-learn, scaler and transformers perform linear and nonlinear transformations, respectively. For example, *MinMaxScaler()* scales all feature belonging to the data set between 0 and 1. Table 12.1 summarizes the main scalers and the transformers available in scikit-learn.

QuantileTransformer() provides nonlinear transformations that shrinks distances between marginal outliers and inliers. Finally, *PowerTransformer()* provides nonlinear transformations in which data are mapped to a normal distribution to stabilize variance and minimize skewness.

Petrelli et al. [118] processed the data set by using *StandardScaler()*. To better understand, see lines 1 and 2 of code Listing 12.2, which show how to apply the *StandardScaler()* to the data displayed in Figs. 12.2 and 12.3.

In addition, Figs. 12.4 and 12.5 show the result of *StandardScaler()* implemented in code Listing 12.2 for the melt and clinopyroxene data, respectively. All features (i.e., the oxides of each chemical element) are characterized by zero mean and unit

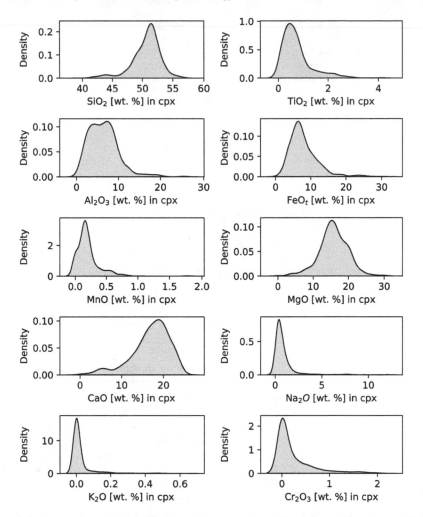

Fig. 12.3 Chemical composition of clinopyroxene phase in training data set of Petrelli et al. [118]

variance. Note that the tree-based methods described in the following section and used here as a ML proxy in geology do not strictly require standardization. However, standardizing helps for data visualization and is useful when applying different methods to the same problem to compare performances with scale-sensitive algorithms, such as support vector machines [97]. Generally speaking, algorithms that depend on measures of distance between predictors require standardization.

```
1 scaler = StandardScaler().fit(X0_train)
2 X_train = scaler.transform(X0_train)
3
4 fig2 = plt.figure(figsize=(8,8))
```

```
 5 for i in range(0,12):
 6     ax3 = fig2.add_subplot(4, 3, i+1)
 7     sns.kdeplot(X_train[:, i],fill=True, color='k',
       facecolor='#ffdfab', ax = ax3)
 8     ax3.set_xlabel('scaled ' + x_labels_melt[i] + '
       the melt')
 9 fig2.align_ylabels()
10 fig2.tight_layout()
11
12 fig3 = plt.figure(figsize=(6,8))
13 for i in range(0,10):
14     ax4 = fig3.add_subplot(5, 2, i+1)
15     sns.kdeplot(X_train[:, i+12],fill=True, color='
       k', facecolor='#ffdfab', ax = ax4)
16     ax4.set_xlabel('scaled ' + x_labels_cpx[i] + '
       in cpx')
17 fig3.align_ylabels()
18 fig3.tight_layout()
```

Listing 12.2 Application of *StandardScaler()* to the data reported in Figs. 12.2 and 12.3

Table 12.1 Scalers and trasformers in Scikit-learn. Descriptions are taken from the official documentation of Scikit-learn

Scaler	Description
sklearn.preprocessing.StandardScaler()	Standardize features by removing the mean and scaling to unit variance (Eq. 12.1)
sklearn.preprocessing.MinMaxScaler()	Transform features by scaling each feature to a given range. The default range is [0, 1]
sklearn.preprocessing.RobustScaler()	Scale features using statistics that are robust against outliers. This scaler removes the median and scales the data according to the quantile range. The default quantile range is the inter-quartile range
Tranformer	**Description**
sklearn.preprocessing.PowerTransformer()	Apply a power transform feature-wise to make data more Gaussian-like. Power transforms are a family of parametric, monotonic transformations that make data more Gaussian-like. As of the writing of this book, PowerTransformer supports the Box-Cox transform and the Yeo-Johnson transform
sklearn.preprocessing.QuantileTransformer()	Transform features using quantile information. This method transforms features to follow a uniform or normal distribution. Therefore, for a given feature, this transformation tends to spread out the most frequent values. It also reduces the impact of (marginal) outliers, making it therefore a robust preprocessing scheme

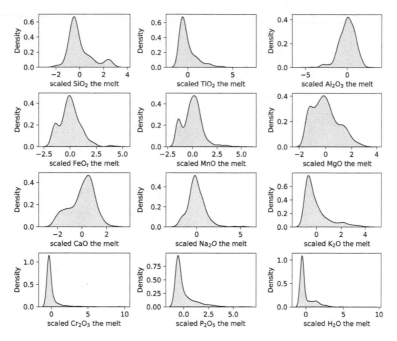

Fig. 12.4 Result of applying *StandardScaler()* to the data reported in Fig. 12.2

Training and testing the model

Similar to the way humans learn from the experience, ML algorithms learn from data. The role of the scaled training data set is to provide the learning experience for the ML algorithm.

Petrelli et al. [118] evaluated numerous ML methods to find the best regressor for problems under investigation. According to their results, the best regressors are Single Decision Trees [89], Random Forest [88], Stochastic Gradient Boosting [94], Extremely Randomized Trees [95], and k-nearest neighbors [85]. How do these regressors work?

Single Decision Trees. A single decision tree model [89] partitions the feature space into a set of regions, and then fits a simple model in each region [128]. To understand how the decision tree model works for a regression problem, consider the example provided by Zhang and Haghani [128], which has a continuous response variable Y and two independent variables X_1 and X_2. The first step of the regression consists of splitting the space defined by X_1 and X_2 into two regions and modeling the response Y (mean of Y) individually in each region. Next, the process continues with each region being split in two until a predetermined stopping rule is satisfied. During each partition, the best fit is achieved through the selection of variables and a split-point [128]. The single tree algorithm forms the base of the random forest, gradient boosting regression, and extremely randomized tree methods. More details on the single decision tree model are available in [89].

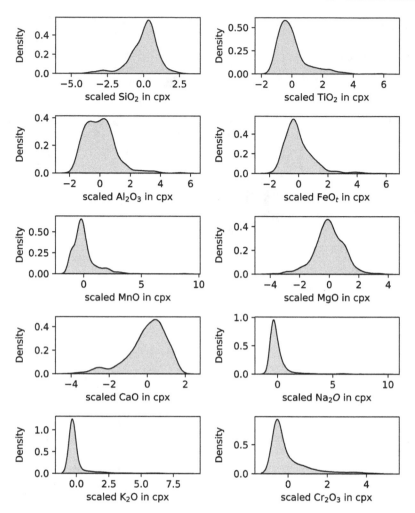

Fig. 12.5 Result of applying *StandardScaler()* to the data reported in Fig. 12.3

Random Forest. The Random Forest algorithm [88] combines two established ML principles [128]: Breiman's "bagging" predictors [87] and the random selection of features [99]. Bagging is a method for producing multiple versions of a predictor and using these to get an aggregated predictor [87]. The multiple versions are created by making bootstrap replicates of the learning set and using these as new learning sets [87]. In the Random Forest algorithm, the bagging predictors generate a diverse subset of data for training-based models [128]. For a given training data set with sample size n, bagging produces k new training sets, each with sample size n, by uniformly sampling from the original training data set with replacement [128]. By sampling with replacement (i.e., bootstrapping), some observations appear more than once in the sample produced, while other observations are left out of the sample [128].

Next, k base models are trained by using the k newly created training sets and coupled by averaging for regression or majority voting for classification [128]. A detailed description of the Random Forest algorithm is available in the literature [88, 111, 128].

Gradient Boosting. In contrast with bagging predictors, the boosting method creates base models sequentially [94, 128]. In the Gradient Boosting algorithm, the prediction capability is progressively improved by sequentially developing multiple models and focusing on the training cases that are difficult to estimate [128]. A key feature in the boosting process is that examples that are hard to estimate using the previous base models appear more often in the training data than those that are correctly estimated [94, 128]. Thus, each successive base model attempts to reduce the errors produced by previous models [128]. A detailed description of the Gradient Boosting algorithm can be found in [94, 128].

Extremely Randomized Trees. The Extremely Randomized Trees algorithm builds an ensemble of regression trees by using the top-down procedure proposed by Geurts et al. [95]. The two main differences compared with other tree-based ensemble methods are that (1) it splits nodes by choosing fully random cut points and (2) it uses the entire learning sample rather than a bootstrapped replica to grow the trees [95]. It has two main parameters: the number of attributes randomly selected at each node and the minimum sample size for splitting a node [95]. It works several times with the (full) original learning sample to generate an ensemble model [95]. The predictions of the trees are aggregated to yield the final prediction—by majority vote in classification problems and by arithmetic averaging in regression problems [95]. A complete description of the Extremely Randomized Trees algorithm is given in [95].

k-nearest neighbors. k-nearest neighbors is a simple algorithm that collects all available cases and predicts the numerical target based on an estimate of similarity, such as distance functions [85]. In detail, it typically uses an inverse-distance-weighted average of the k-nearest neighbors [85]. The weight of each training instance can be uniform or computed by using a kernel function, which could depend on the distance (as opposed to the similarity) between itself and the test instance. Note that the prediction using a single neighbor is just the target value of the nearest neighbor [85]. The Euclidean distance metric is commonly used to measure the distance between two instances. A detailed description of the k-nearest neighbors algorithm is available in [85].

Table 12.2 lists the scikit-learn implementation of Single Decision Trees [89], Random Forests [88], Stochastic Gradient Boosting [94], Extremely Randomized Trees [95], and k-nearest neighbors [85].

The training and test processes can be easily done by using scikit-learn, as shown in code Listing 12.3, which proceeds in the following steps:

1. define and train the algorithm (in our case, the Extremely Randomized Trees method, see lines 2 and 5);
2. import the test data set and scale it in accordance with the rules used for the train data set (lines 8–17);
3. predict the test data set (line 20);

Table 12.2 ML regressors reported by Petrelli et al. [118]

Algorithm	scikit-learn
Single Decision Trees	class sklearn.tree.**DecisionTreeRegressor**()
Random Forest	class sklearn.ensemble.**RandomForestRegressor**()
Gradient Boosting	class klearn.ensemble.**GradientBoostingRegressor**()
Extremely Randomized Trees	class sklearn.ensemble.**ExtraTreesRegressor**()
k-nearest neighbors	class sklearn.neighbors.**KNeighborsRegressor**()

4. select one or more metrics to evaluate the results (lines 23 and 24);
5. evaluate the results (lines 27–33) shown in Fig. 12.6.

```
 1 # Define the regressor, in our case the Extra Tree
       Regressor
 2 regr = ExtraTreesRegressor(n_estimators=550,
       criterion='mse', max_features=22, random_state
       =280) # random_state fixed for reproducibility
 3
 4 # Train the model
 5 regr.fit(X_train, Y_train.ravel())
 6
 7 # Import the test data set
 8 my_test_dataset = pd.read_excel('
       GlobalDataset_Final_rev9_Test.xlsx', usecols =
       "A:M,O:X,Z:AA", skiprows=1, engine='openpyxl')
 9 my_test_dataset.columns = [c.replace('.1', 'cpx')
       for c in my_test_dataset.columns]
10 my_test_dataset = my_test_dataset.fillna(0)
11
12 X0_test = my_test_dataset.iloc[:, 1:23]
13 Y_test= np.array([my_test_dataset.T_K]).T
14 labels_test = np.array([my_test_dataset.Sample_ID])
       .T
15
16 # Scale the test dataset
17 X_test_scaled = scaler.transform(X0_test)
18
19 # Make the prediction on the test data set
20 predicted = regr.predict(X_test_scaled)
21
22 # Evaluate the results using the R2 and RMSE
23 r2 = r2_score(Y_test, predicted)
24 rmse = np.sqrt(mean_squared_error(predicted, Y_test
       ))
25
26 # Plot data
```

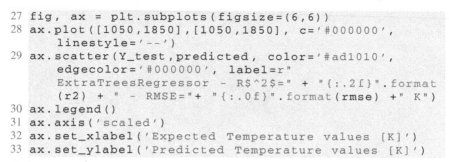

```
27 fig, ax = plt.subplots(figsize=(6,6))
28 ax.plot([1050,1850],[1050,1850], c='#000000',
      linestyle='--')
29 ax.scatter(Y_test,predicted, color='#ad1010',
      edgecolor='#000000', label=r"
      ExtraTreesRegressor - R$^2$=" + "{:.2f}".format
      (r2) + " - RMSE="+ "{:.0f}".format(rmse) +" K")
30 ax.legend()
31 ax.axis('scaled')
32 ax.set_xlabel('Expected Temperature values [K]')
33 ax.set_ylabel('Predicted Temperature values [K]')
```

Listing 12.3 Training and testing the *ExtraTreesRegressor()* algorithm to predict temperature

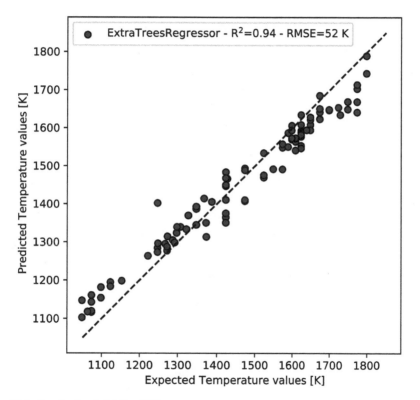

Fig. 12.6 Result of code Listing 12.3

Appendix A
Python Packages and Resources for Geologists

Python is a widely-used programming language and the developers involved in Earth Science have created many libraries and packages to solve Geology problems. As an example, in the Awesome Open Geoscience repository[1] and in a blog post by the American Geophysical Union Hydrology Section Student Subcommittee (AGU-H3S)[2] they list many Python projects, also dividing them by application fields. Also, the book titled "Pythonic Geodynamics" (Morra, 2018) lists several examples of the application of Python programming to geodynamic modeling. Finally, you should look at Pangeo,[3] a community of researchers who work together to develop a platform for Big Data geoscience.

A.1 Python Libraries for Geologists

I reported a list of Python libraries that have been developed for Earth Scientists in the book repository.[4]

A.2 Python Learning Resources for Geologists

Surfing the World Wide Web, Geologists can find many excellent resources to improve their knowledge and abilities in Python programming. As an example, the

---

[1] https://github.com/softwareunderground/awesome-open-geoscience.

[2] https://agu-h3s.org/2021/03/29/resources-for-programming-in-hydrology/.

[3] https://pangeo.io/.

[4] https://github.com/petrelli-m/python_earth_science_book.

© The Editor(s) (if applicable) and The Author(s), under exclusive license to Springer Nature Switzerland AG 2021
M. Petrelli, *Introduction to Python in Earth Science Data Analysis*,
Springer Textbooks in Earth Sciences, Geography and Environment,
https://doi.org/10.1007/978-3-030-78055-5

Earth Lab[5] at the University of Colorado (Boulder) provides many tutorials and course lessons about the application of Python methods and techniques to Earth Sciences problems.

Many other Universities (e.g., The University of Melbourne,[6] The University of Bergen,[7] the Max Planck Institute for Meteorology,[8] The Australian National University,[9] and The University of Perugia,[10] to cite a few) have active courses (May 2021) teaching the application of Python programming to Earth Scientists.

Also, various active researchers continuously spread excellent material to improve the application of advanced statistical and computational techniques in Python to geologists. I reported some examples in the book repository.

Morra, G. (2018). *Pythonic Geodynamics: Implementations for Fast Computing.* Springer International Publishing, ISBN 978-3-319-55682-6.

---

[5] https://earthlab.colorado.edu.

[6] https://handbook.unimelb.edu.au/2017/subjects/erth90051.

[7] https://www.uib.no/en/course/GEOV302.

[8] https://bit.ly/3xeLAAQ.

[9] https://programsandcourses.anu.edu.au/2019/course/emsc8033.

[10] https://www.unipg.it/personale/maurizio.petrelli/en/teaching.

Appendix B
Introduction to Object Oriented Programming

B.1 Object-Oriented Programming

Definition: "Object-oriented programming (OOP) is a programming paradigm based on the concept of objects, which can contain data and code. The data take the form of fields (often known as attributes or properties), and code takes the form of procedures (often known as methods)."[11]

The main building blocks of OOP are classes and objects.

Typically, classes represent broad categories, like the items of an online shop or physical objects that share similar attributes. All objects created from a specific class share the same attributes (e.g., color, size, etc.). In practice, a class is the blueprint, whereas an object contains real data and is built from a class. The creation of a new object from a class is called "instantiating" an object.

For example, we could define a class for the items of an online shop containing the following attributes: color, size, description, and price. We could then instantiate numerous objects, each characterized by a specific color, size, description, and price. When we define a Dataframe or a Figure in Python, we are creating objects using the class *pandas.DataFrame()* and *matplotlib.figure.Figure()*.

Classes also contain functions, which are called methods in OOP. Methods are defined within the class and perform actions or computations using the data contained within the given object. For example, *.var()* and *.mean()* are methods available for the objects belonging to the class *pandas.DataFrame()*.

---

[11] https://en.wikipedia.org/wiki/Object-oriented_programming.

© The Editor(s) (if applicable) and The Author(s), under exclusive license to Springer Nature Switzerland AG 2021
M. Petrelli, *Introduction to Python in Earth Science Data Analysis*,
Springer Textbooks in Earth Sciences, Geography and Environment,
https://doi.org/10.1007/978-3-030-78055-5

B.2   Defining Classes, Attributes, and Methods in Python

The class statement followed by a unique class name and a colon define a class.
By convention, Python class names are written in capitalized words (e.g., MyClass).
For example, the code Listing B.1 defines a class named Circle after importing the
NumPy library that will be required in the subsequent development of the class.

```
1   import numpy as np
2
3   class Circle:
4       # Attributes an methods here
```
Listing B.1 Defining a new class in Python

The attributes of the class are defined in the method called .__init__(), which can
contain many parameters. However, the first parameters is always a variable called
"self."
   For example, in code Listing B.2, we define the attribute radius for the class Circle.

```
1   import numpy as np
2
3   class Circle:
4
5       def __init__(self, radius):
6           self.radius = radius
```
Listing B.2 Adding attributes to a class

```
1    import numpy as np
2
3    class Circle:
4
5        def __init__(self, radius):
6            self.radius = radius
7
8        # my first Instance method
9        def description(self):
10           return "circle with radius equal to {:.2f}".format(self.radius
             )
11
12       # my secong instance method
13       def area(self):
14           return np.pi * self.radius ** 2
15
16       # my secong instance method
17       def circumference(self):
18           return 2 * np.pi * self.radius
19
20       # my tird instance method
21       def diameter(self):
22           return 2 * np.pi
```
Listing B.3 Adding methods to a class

Finally, methods are functions that are defined inside a class and can only be called from an object belonging to that specific class. The code Listing B.3 implements the methods *description()*, *area()*, *circumference()*, and *diameter()*.

Finally, code Listing B.4 shows how to create (i.e., instantiate) a new Circle object called my_Circle, print its description, and calculate its area.

```python
import numpy as np

class Circle:

    def __init__(self, radius):
        self.radius = radius

    # my first Instance method
    def description(self):
        return "circle with radius equal to {:.2f}".format(self.radius
        )

    # my secong instance method
    def area(self):
        return np.pi * self.radius ** 2

    # my secong instance method
    def circumference(self):
        return 2 * np.pi * self.radius

    # my tird instance method
    def diameter(self):
        return 2 * np.pi


my_Circle = Circle(radius=2)

# Description
print(my_Circle.description())

# Calculate and report the area
my_Area = my_Circle.area()

# Reporting the area of my_Circle
print("The area of a {} is equal to {:.2f}".format(my_Circle.
    description(), my_Area))
```

Listing B.4 Instantiating an object of the Circle class and using its methods (i.e., functions)

Appendix C
The Matplotlib Object Oriented API

C.1 Matplotlib Application Programming Interfaces

As reported in Sect. 3.1, there are two main Application Programming Interfaces (APIs) to use Matplotlib:

OO-style: Using the OO-style, you explicitly define the objects governing the content and the aesthetics of a diagram (i.e., figures and axes) and call methods on them to create your diagram.

pyplot style: This is the simplest way of plotting in matplotlib. Using the pyplot style, you rely on pyplot to automatically create and manage the objects governing your diagram. You also use pyplot functions for plotting.

Regarding the use of a specific style, the official documentation of matplolib states (Feb, 2021) "Matplotlib's documentation and examples use both the OO and the pyplot approaches (which are equally powerful), and you should feel free to use either (however, it is preferable pick one of them and stick to it, instead of mixing them). In general, we suggest to restrict pyplot to interactive plotting (e.g., in a Jupyter notebook), and to prefer the OO-style for non-interactive plotting (in functions and scripts that are intended to be reused as part of a larger project)."[12]

C.2 Matplotlib Object Oriented API

As reported is Sect. 1.2 and in Appendix B, when using the OOP paradigm, everything is an object instantiated from a class. The following descriptions are taken and adapted from the matplotlib official documentation[13]:

---

[12] https://matplotlib.org/stable/tutorials/introductory/usage.html.

[13] https://matplotlib.org.

© The Editor(s) (if applicable) and The Author(s), under exclusive license to Springer Nature Switzerland AG 2021
M. Petrelli, *Introduction to Python in Earth Science Data Analysis*,
Springer Textbooks in Earth Sciences, Geography and Environment,
https://doi.org/10.1007/978-3-030-78055-5

215

The main classes governing a diagram in matplotlib are listed below.

Figure. The Figure object embeds the whole diagram and keeps track of all child axes, art (e.g., titles, figure legends, etc.), and the canvas. A Figure can contain any number of Axes, but will typically have at least one.

Axes. Axes are what you typically think of when using the word "plot." It is the region of the Figure where you plot your data. A given Figure can host many Axes, but a specific Axes object can be in only a single Figure.

Axis: The Axis takes care of setting the graph limits and generating the ticks (i.e., the marks on the axis) and ticklabels (i.e., strings labeling the ticks). The location of the ticks is determined by an object called Locator, and the ticklabel strings are formatted by a Formatter. Tuning the Locator and Formatter gives very fine control over tick locations and labels. Data limits can be also controlled via the Axes. *Axes.set_xlim()* and *axes.Axes.set_ylim()* methods). Each Axes has a title (set via *set_title()*), an *x* label set via *set_xlabel()* and a *y* label set via *set_ylabel()*. Note that Axes and Axis are two different type of objects in matplotlib.

Artists: An artist is any object that you can see within a Figure. Artists includes Text objects, Line2D objects, collections objects, and many others. When a Figure is rendered, all Artists are drawn on the canvas.

C.3   Fine Tuning Geological Diagrams Using the OO-Style

Using the OO-style, we can access any class in matplotlib. These classes provide numerous attributes and methods to fine tune a geological diagram.

For example, code Listing C.1 highlights, with embedded referencing to the official documentation, how to further personalize a geological diagram we developed in this book (Fig. 4.2). Fine tuning this diagram further improves the quality of our artwork. Figure C.1 shows the result of code Listing C.1.

```
1  import matplotlib.pyplot as plt
2  import matplotlib as mpl
3  from matplotlib.ticker import MultipleLocator, FormatStrFormatter,
       AutoMinorLocator
4  import pandas as pd
5  import numpy as np
6
7  myDataset = pd.read_excel('Smith_glass_post_NYT_data.xlsx', sheet_name
       ='Supp_traces')
8
9  fig, ax = plt.subplots()
10 # Figure managment
11 # https://matplotlib.org/stable/api/_as_gen/matplotlib.figure.Figure.
       html
12
13 # Axes managment: https://matplotlib.org/stable/api/axes_api.html
14
15 # select your style
```

```
16  #   https://matplotlib.org/stable/gallery/style_sheets/
        style_sheets_reference.html
17  mpl.style.use('ggplot')
18
19  # Make the plot
20  ax.hist(myDataset.Zr, density=True, bins='auto', color='Tab:blue',
        edgecolor='k', alpha=0.8, label = 'CFC recent activity')
21
22  # Commonnly used personalizations
23  ax.set_xlabel('Zr [ppm]')
24  ax.set_ylabel('Probability density')
25  ax.set_title('Zr sample distribution')
26  ax.set_xlim(-100, 1100)
27  ax.set_ylim(0,0.0055)
28  ax.set_xlabel(r'Zr [$\mu \cdot g^{-1}$]')
29  ax.set_ylabel('Probability density')
30  ax.set_xticks(np.arange(0, 1100 + 1, 250))   # adjust the x tick
31  ax.set_yticks(np.arange(0, 0.0051, .001))   # adjust the y tick
32
33  # Major and minor ticks
34  # https://matplotlib.org/stable/gallery/ticks_and_spines/
        major_minor_demo.html
35
36  ax.xaxis.set_minor_locator(AutoMinorLocator())
37  ax.tick_params(which='both', width=1)
38  ax.tick_params(which='major', length=7)
39  ax.tick_params(which='minor', length=4)
40
41  ax.yaxis.set_minor_locator(MultipleLocator(0.0005))
42  ax.tick_params(which='both', width=1)
43  ax.tick_params(which='major', length=7)
44  ax.tick_params(which='minor', length=4)
45
46  # Spine management
47  # https://matplotlib.org/stable/api/spines_api.html
48
49  ax.spines["top"].set_color("#363636")
50  ax.spines["right"].set_color("#363636")
51  ax.spines["left"].set_color("#363636")
52  ax.spines["bottom"].set_color("#363636")
53
54  # Spine placement
55  # https://matplotlib.org/stable/gallery/ticks_and_spines/
        spine_placement_demo.html
56
57  # Advanced Annotations
58  # https://matplotlib.org/stable/tutorials/text/annotations.html#
        plotting-guide-annotation
59  ax.annotate("Mean Value",
60              xy=(myDataset.Zr.mean(), 0.0026), xycoords='data',
61              xytext=(myDataset.Zr.mean() + 250, 0.0035), textcoords='
        data',
62              arrowprops=dict(arrowstyle="fancy",
63                              color="0.5",
64                              shrinkB=5,
65                              connectionstyle="arc3,rad=0.3",
66                              ),
67              )
68
69  ax.annotate("Modal \n value ",
70              xy=(294, 0.0045), xycoords='data',
71              xytext=(0, 0.005), textcoords='data',
```

```
72              arrowprops=dict(arrowstyle="fancy",
73                              color="0.5",
74                              shrinkB=5,
75                              connectionstyle="arc3,rad=-0.3",
76                              ),
77          )
78
79  # Legend: https://matplotlib.org/stable/api/legend_api.html
80  ax.legend(title = 'My Legend')
81
82  fig.tight_layout()
```

Listing C.1 Advanced personalization of matplotlib diagrams

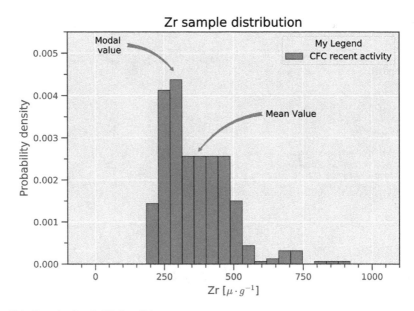

Fig. C.1 Result of code Listing C.1

Appendix D
Working with Pandas

D.1 How to Perform Common Operations in Pandas

Importing an Excel file:

```
1  In [1]: import pandas as pd
2
3  In [1]: myDataset = pd.read_excel('Smith_glass_post_NYT_data.xlsx'
       , sheet_name='Supp_traces')
```

Importing a .csv file:

```
1  In [1]: import pandas as pd
2
3  In [1]: myDataset = MyData = pd.read_csv('DEM.csv')
```

Get the column labels:

```
1  In [3]: myDataset.columns
2  Out[3]: Index(['Analysis no.', 'Strat. Pos.', 'Eruption', '
       controlcode', 'Sample', 'Epoch', 'Crater size', 'Date of
       analysis', 'Si/bulk cps', 'SiO2* (EMP)', 'Sc', 'Rb', 'Sr', 'Y'
       , 'Zr', 'Nb', 'Cs', 'Ba', 'La', 'Ce', 'Pr', 'Nd', 'Sm', 'Eu',
       'Gd', 'Tb', 'Dy', 'Ho', 'Er', 'Tm', 'Yb', 'Lu', 'Hf', 'Ta', '
       Pb', 'Th', 'U'],dtype='object')
```

Get the shape (i.e., height and width) of a DataFrame:

```
1  In [4]: myDataset.shape
2  Out[4]: (370, 37)
```

© The Editor(s) (if applicable) and The Author(s), under exclusive license to Springer
Nature Switzerland AG 2021
M. Petrelli, *Introduction to Python in Earth Science Data Analysis*,
Springer Textbooks in Earth Sciences, Geography and Environment,
https://doi.org/10.1007/978-3-030-78055-5

Select a single column:

```
 1  In[5]: myDataset['Rb']
 2  Out[5]:
 3  0         355.617073
 4  1         367.233701
 5  2         293.320592
 6  3         344.871192
 7  4         352.352196
 8             ...
 9  365       358.479709
10  366       405.655463
11  367       328.080366
12  368       333.859656
13  369       351.240272
14  Name: Rb, Length: 370, dtype: float64
```

or

```
 1  In[6]: myDataset.Rb
 2  Out[6]:
 3  0         355.617073
 4  1         367.233701
 5  2         293.320592
 6  3         344.871192
 7  4         352.352196
 8             ...
 9  365       358.479709
10  366       405.655463
11  367       328.080366
12  368       333.859656
13  369       351.240272
14  Name: Rb, Length: 370, dtype: float64
```

Select the first two rows of the whole DataFrame:

```
 1  In[7]: myDataset[0:2]
 2  Out[7]:
 3  Analysis no.   ...           Pb          Th           U
 4  0              ...     60.930984   35.016435    9.203411
 5  1              ...     59.892427   34.462577   10.459280
 6  [2 rows x 37 columns]
```

Select the first four rows of a single column:

```
 1  In[8]: myDataset['Rb'][0:4]
 2  Out[8]:
 3  0         355.617073
 4  1         367.233701
 5  2         293.320592
 6  3         344.871192
 7  Name: Rb, dtype: float64
```

Convert the first four rows of a single column to a NumPy array:

```
 1  Out[9]: myDataset.Rb[0:4].to_numpy()
 2  Out[9]: array([355.61707274, 367.23370121, 293.32059158,
        344.87119168])
```

Select a single cell:

```
1  In[10]: myDataset['Rb'][4]
2  Out[10]: 352.3521959503882
```

or, use row an column indexes (note that rows and columns are reversed with respect to the previous example):

```
1  In[11]: myDataset.iloc[4,11]
2  Out[11]: 352.3521959503882
```

Sort:

```
1  In[12]: myDataset.sort_values(by='SiO2* (EMP)', ascending=False)
2  Out[12]:
3  Analysis no.  ...    SiO2* (EMP) ...      Th            U
4  228           ...    62.410000   ...      56.114101   15.548608
5  236           ...    62.410000   ...      47.402098   12.345041
6  ...           ...    ...         ...      ...          ...
7  304           ...    54.425402   ...      16.539421    5.256582
8  318           ...    54.425402   ...      16.539421    5.256582
9  [370 rows x 37 columns]
```

Filter:
(1) define a sub DataFrame containing all the samples with zirconium above 400

```
1  In[13]: myDataset1 = myDataset[myDataset.Zr > 400]
```

(2) define a sub DataFrame containing all the samples with zirconium between 400 and 450

```
1  In[14]: myDataset2 = myDataset[((myDataset.Zr > 400)&(myDataset.Zr
          < 500))]
```

Managing missing data:
(1) drop any rows that have missing data

```
1  In[15]: myDataset3 = myDataset.dropna(how='any')
2  In[16]: myDataset.shape
3  Out[16]: (370, 37) <- the original data set
4  In[17]: myDataset3.shape
5  Out[17]: (366, 37) <- 4 samples contained missing data
```

(2) replace missing data with a fixed value (e.g., 5)

```
1  In[18]: myDataset4 = myDataset.fillna(value=5)
```

Further Readings

Part I: Python for Geologists, a Kick-Off

1. Beazley D, Jones B (2013) Python Cookbook. O'Reilly Media, Inc
2. Bisong E (2019) Matplotlib and Seaborn. In: Building machine learning and deep learning models on google cloud platform, pp 151–165. Apress
3. Bressert E (2012) SciPy and NumPy: an overview for developers. O'Reilly Media, Inc
4. Chen DY (2017) Pandas for everyone: Python data analysis. Addison-Wesley Professional
5. Dowek G, Lévy J-J (2011) Introduction to the theory of programming languages. Springer, London. https://doi.org/10.1007/978-0-85729-076-2
6. Downey A (2016) Think Python. O'Reilly Media, Inc
7. Gabbrielli M, Martini S (2010) Programming languages: principles and paradigms. Springer, London. https://doi.org/10.1007/978-1-84882-914-5
8. Hunt J (2019) A beginners guide to Python 3 programming. Springer International Publishing. https://doi.org/10.1007/978-3-030-20290-3
9. Lubanovic B (2019) Introducing Python: modern computing in simple packages. O'Reilly Media, Inc
10. Matthes E (2019) Python crash course: a hands-on, project-based introduction to programming. No Starch Press
11. McKinney W (2017) Python for Data Analysis, 2nd Edition [Book]. O'Reilly Media, Inc
12. Meurer A, Smith CP, Paprocki M, Čertík O, Kirpichev SB, Rocklin M, Kumar A, Ivanov S, Moore JK, Singh S, Rathnayake T, Vig S, Granger BE, Muller RP, Bonazzi F, Gupta H, Vats S, Johansson F, Pedregosa F, Scopatz A (2017) SymPy: symbolic computing in Python. PeerJ Comput Scie 3:e103. https://doi.org/10.7717/peerj-cs.103
13. Paper D (2020) Hands-on Scikit-Learn for machine learning applications. Apress. https://doi.org/10.1007/978-1-4842-5373-1
14. Rollinson H (1993) Using geochemical data: evaluation. Presentation, Routledge, Interpretation. Routledge
15. Rossant C (2018) IPython Cookbook, 2nd edn. Packt Publishing
16. Smith V, Isaia R, Pearce N (2011) Tephrostratigraphy and glass compositions of post-15 kyr Campi Flegrei eruptions: implications for eruption history and chronostratigraphic markers. Quat Sci Revi 30(25–26):3638–3660. https://doi.org/10.1016/J.QUASCIREV.2011.07.012

© The Editor(s) (if applicable) and The Author(s), under exclusive license to Springer 223
Nature Switzerland AG 2021
M. Petrelli, *Introduction to Python in Earth Science Data Analysis*,
Springer Textbooks in Earth Sciences, Geography and Environment,
https://doi.org/10.1007/978-3-030-78055-5

17. Sweigart A (2019) Automate the boring stuff with Python: practical programming for total beginners. No Starch Press
18. Turbak FA, Gifford DK (2008) Design concepts in programming languages. MIT Press
19. Van Roy P, Haridi S (2004) Concepts, techniques, and models of computer programming. MIT Press

Part II: Describing Geological Data

20. Blundy J, Wood B (1994) Prediction of crystal-melt partition coefficients from elastic moduli. Nature 372(6505):452–454. https://doi.org/10.1038/372452a0
21. Blundy J, Wood B (2003) Partitioning of trace elements between crystals and melts. Earth Planet Sci Lett 210:383–397
22. Branch MA, Coleman TF, Li Y (1999) A subspace, interior, and conjugate gradient method for large-scale bound-constrained minimization problems. SIAM J Sci Comput 21(1):1–23. https://doi.org/10.1137/S1064827595289108
23. Chatterjee S, Hadi AS (2013) Regression analysis by example, 5th Edn. Wiley
24. Healy K (2019) Data visualization: a practical introduction. Princeton University Press
25. Heumann C, Schomaker M, Shalabh. (2017) Introduction to statistics and data analysis: With exercises, solutions and applications in R. Springer International Publishing. https://doi.org/10.1007/978-3-319-46162-5
26. Holcomb ZC (1998) Fundamentals of descriptive statistics, 1st Edn. Routledge
27. Kopka H, Daly PW (2003) Guide to LaTeX. Addison-Wesley Professional
28. Lamport L (1994) LaTeX: A document preparation system. Addison Wesley
29. Meltzer A, Kessel R (2020) Modelling garnet-fluid partitioning in H2O-bearing systems: a preliminary statistical attempt to extend the crystal lattice-strain theory to hydrous systems. Contribut. Mineral. Petrol. 175(8):80. https://doi.org/10.1007/s00410-020-01719-8
30. Mittelbach F, Goossens M, Braams J, Carlisle D, Rowley C (2004) The LaTeX Companion, 2nd edn. Addison-Wesley Professional
31. Montgomery DC, Peck EA, Geoffrey Vining G (2012) Introduction to linear regression analysis. Wiley
32. Moré JJ (1978) The Levenberg–Marquardt algorithm: implementation and theory. In G Watson (ed), Numerical analysis. lecture notes in mathematics, pp 105–116. Springer, Berlin, Heidelberg. https://doi.org/10.1007/BFb0067700
33. Motulsky H, Christopoulos A (2004) Fitting models to biological data using linear and nonlinear regression: a practical guide to curve fitting. Oxford University Press
34. Olson DL, Lauhoff G (2019) Descriptive data mining. Springer, Singapore. https://doi.org/10.1007/978-981-13-7181-3
35. Ross SM (2017) Introductory statistics, 4th edn. Academic Press
36. Seber GAF, Wild CJ (1989). Nonlinear regression. Wiley
37. Tufte E (2001) The visual display of quantitative information, 2nd edn. Graphics Press
38. Voglis C, Lagaris IE (2004) A rectangular trust region dogleg approach for unconstrained and bound constrained nonlinear optimization. In: Simos T, Maroulis G (eds), Wseas international conference on applied mathematics, corfu, greece, pp 562–565. Taylor, Francis Inc. https://doi.org/10.1201/9780429081385-138

Part III: Integrals and Differential Equations in Geology

39. Agarwal RP, O'Regan D (2008) An introduction to ordinary differential equations. Springer
40. Anderson DL (1989) Theory of the Earth. Blackwell Scientific Publications
41. Atkinson KE, Han W, Stewart D (2009) Numerical solution of ordinary differential equations. Wiley
42. Burd A (2019) Mathematical methods in the Earth and environmental sciences. Cambridge University Press
43. Costa F, Chakraborty S, Dohmen R (2003) Diffusion coupling between trace and major elements and a model for calculation of magma residence times using plagioclase. Geochimica et Cosmochimica Acta 67(12):2189–2200. https://doi.org/10.1016/S0016-7037(02)01345-5
44. Costa F, Shea T, Ubide T (2020) Diffusion chronometry and the timescales of magmatic processes. Nat Rev Earth Environ 1(4):201–214. https://doi.org/10.1038/s43017-020-0038-x
45. Crank J (1975) The mathematics of diffusion, 2nd edn. Clarendon Press
46. Dziewonski AM, Anderson DL (1981) Preliminary reference Earth model. Phys Earth Planet Inter 25(4):297–356. https://doi.org/10.1016/0031-9201(81)90046-7
47. Fick A (1855) Ueber Diffusion. Annalen der Physik und Chemie 170(1):59–86. https://doi.org/10.1002/andp.18551700105
48. Griffiths DF, Higham DJ (2010) Numerical methods for ordinary differential equations. Springer, London
49. King D, Billingham J, Otto SR (2003) Differential equations. Linear, non- linear, ordinary, partial. Cambridge University Press
50. Li Z, Qiao Z, Tang T (2017) Numerical solution of differential equations. Cambridge Univ Press. https://doi.org/10.1017/9781316678725
51. Linge S, Langtangen HP (2017) Finite difference computing with PDEs, 1st edn. Springer International Publishing. https://doi.org/10.1007/978-3-319-55456-3
52. Mazumder S (2015) Numerical methods for partial differential equations: finite difference and finite volume methods. Academic press
53. Moore A, Coogan L, Costa F, Perfit M (2014) Primitive melt replenishment and crystal-mush disaggregation in the weeks preceding the 2005–2006 eruption 9° 50' N, EPR. Earth Planet Sci Lett 403:15–26. https://doi.org/10.1016/J.EPSL.2014.06.015
54. Morton KW, Mayers DF (2005) Numerical solution of partial differential equations. Cambridge Univ Press. https://doi.org/10.1017/CBO9780511812248
55. Paul A, Laurila T, Vuorinen V, Divinski SV (2014) Thermodynamics, diffusion and the kirkendall effect in solids. Springer Int Publ. https://doi.org/10.1007/978-3-319-07461-0
56. Poirier J (2000) Introduction to the physics of the earth's interior, 2 edn. Cambridge University Press
57. Priestley H (1997) Introduction to integration. Oxford University Press
58. Slavinić P, Cvetković Marko (2016) Volume calculation of subsurface structures and traps in hydrocarbon exploration—a comparison between numerical integration and cell based models. Open Geosci 8(1). https://doi.org/10.1515/geo-2016-0003
59. Strang G, Herman E, OpenStax College, Open Textbook Library (2016) Calculus. Volume 1. OpenStax—Rice University
60. Zill D (2012) A first course in differential equations with modeling applications. Cengage Learning, Inc

Part IV: Probability Density Functions and Error Analysis

61. Agterberg F (2018) Statistical modeling of regional and worldwide size- frequency distributions of metal deposits. Handbook of mathematical geosciences, pp 505–523. Springer
62. Ahrens LH (1953) A fundamental law of geochemistry. Nature 172(4390):1148. https://doi.org/10.1038/1721148a0
63. Ballio F, Guadagnini A (2004) Convergence assessment of numerical Monte Carlo simulations in groundwater hydrology. Water Res Res 40(4):4603. https://doi.org/10.1029/2003WR002876
64. Barbu A, Zhu S-C (2020) Monte carlo methods, 1st edn. Springer Nature. Davies J, Marzoli A, Bertrand H, Youbi N, Ernesto M, Schaltegger U (2017) End-Triassic mass extinction started by intrusive CAMP activ- ity. Nat Communic 8(1):15596. https://doi.org/10.1038/ncomms15596
65. Everitt B (2006) The Cambridge dictionary of statistics, 3rd edn. Cambridge University Press
66. Gramacki A (2018) Nonparametric Kernel density estimation and its computational aspects, vol 37. Springer International Publishing
67. Haramoto H, Matsumoto M, L'Ecuyer P (2008) A fast jump ahead algorithm for linear recurrences in a polynomial space. Sequences and their applications—seta 2008, pp 290–298. Springer
68. Hughes I, Hase T (2010) Measurements and their uncertainties: a practical guide to modern error analysis. Oxford University Press
69. Johnston D (2018) Random number generators—principles and practices: a guide for engineers and programmers. Walter de Gruyter GmbH
70. Liu SA, Wu H, Shen SZ, Jiang G, Zhang S, Lv Y, Zhang H, Li S (2017) Zinc isotope evidence for intensive magmatism immediately before the end-Permian mass extinction. Geology 45(4):343–346. https://doi.org/10.1130/G38644.1
71. O'Neill ME (2014) PCG: a family of simple fast space-efficient statistically good algorithms for random number generation. www.pcg-random.org
72. Puetz SJ (2018) A relational database of global U-Pb ages. Geosci Front 9(3):877–891. https://doi.org/10.1016/J.GSF.2017.12.004
73. Reimann C, Filzmoser P (2000) Normal and lognormal data distribution in geochemistry: death of a myth. Consequences for the statistical treatment of geochemical and environmental data. Environ Geol 39(9):1001–1014. https://doi.org/10.1007/s002549900081
74. Rocholl A (1998) Major and trace element composition and homogeneity of microbeam reference material: basalt glass USGS BCR-2G. Geostand Geoanal Res 22(1):33–45. https://doi.org/10.1111/j.1751-908X.1998.tb00543.x
75. Salmon JK, Moraes MA, Dror RO, Shaw DE (2011) Parallel random numbers: as easy as 1, 2, 3. In: SC'11: Proceedings of 2011 international conference for high performance computing, networking, storage and analysis, pp 1–12. https://doi.org/10.1145/2063384.2063405
76. Schwartz LM (1975) Random error propagation by Monte Carlo simulation. Anal Chem 47(6):963–964. https://doi.org/10.1021/ac60356a027
77. Silverman BW (1998) Density estimation for statistics and data analysis. Hall/CRC, Chapman
78. Taylor JR (1997) An introduction to error analysis: the study of uncertainties in physical measurements, 2nd edn. University Science Books
79. Tegner C, Marzoli A, McDonald I, Youbi N, Lindström S (2020) Platinum- group elements link the end-Triassic mass extinction and the Central Atlantic Magmatic Province. Sci Rep 10(1):1–8. https://doi.org/10.1038/s41598-020-60483-8
80. Tobutt DC (1982) Monte Carlo Simulation methods for slope stability. Comput Geosci 8(2):199–208. https://doi.org/10.1016/0098-3004(82)90021-8
81. Troyan V, Kiselev Y (2010) Statistical methods of geophysical data processing. World Scientific Publishing Company
82. Ulianov A, Müntener O, Schaltegger U (2015) The ICPMS signal as a Poisson process: a review of basic concepts. J Anal Atom Spectr 30(6):1297–1321. https://doi.org/10.1039/C4JA00319E

83. Wang Z, Yin Z, Caers J, Zuo R (2020) A Monte Carlo-based framework for risk-return analysis in mineral prospectivity mapping. Geosci Front 11(6):2297–2308. https://doi.org/10.1016/j. gsf.2020.02.010

Part V: Robust Statistics and Machine Learning

84. Abedi M, Norouzi G-H, Bahroudi A (2012) Support vector machine for multi- classification of mineral prospectivity areas. Comput Geosci 46:272–283. https://doi.org/10.1016/j.cageo. 2011.12.014
85. Bentley JL (1975) Multidimensional binary search trees used for associative searching. Commun ACM 18(9):509–517. https://doi.org/10.1145/361002.361007
86. Bishop C (2006) Pattern recognition and machine learning. Springer
87. Breiman L (1996) Bagging predictors. Mach Learn 24(2):123–140. https://doi.org/10.1023/ A:1018054314350
88. Breiman L (2001) Random forests. Mach Learn 45(1):5–32. https://doi.org/10.1023/A: 1010933404324
89. Breiman L, Friedman JH, Olshen RA, Stone CJ (2017) Classification and regression trees. CRC Press. https://doi.org/10.1201/9781315139470
90. Cannata A, Montalto P, Aliotta M, Cassisi C, Pulvirenti A, Privitera E, Patanè D (2011) Clustering and classification of infrasonic events at Mount Etna using pattern recognition techniques. Geophys J Int 185(1):253–264. https://doi.org/10.1111/j.1365-246X.2011.04951.x
91. D'Agostino R, Pearson ES (1973) Tests for departure from normality. Biometrika 60:613–622
92. D'Agostino RB (1971) An omnibus test of normality for moderate and large sample size. Biometrika 58:341–348
93. Devine JD, Murphy MD, Rutherford MJ, Barclay J, Sparks RSJ, Carroll MR, Young SR, Gardner IE (1998) Petrologic evidence for pre-eruptive pressure-temperature conditions, and recent reheating, of andesitic magma erupting at the Soufriere Hills Volcano, Montserrat, W.I. Geophys Res Lett 25(19):3669–3672. https://doi.org/10.1029/98GL01330
94. Friedman JH (2002) Stochastic gradient boosting. Comput Stat Data Anal 38(4):367–378. https://doi.org/10.1016/S0167-9473(01)00065-2
95. Geurts P, Ernst D, Wehenkel L (2006) Extremely randomized trees. Mach Learn 63(1):3–42. https://doi.org/10.1007/s10994-006-6226-1
96. Goldstein E, Coco G (2014) A machine learning approach for the prediction of settling velocity. Water Res Res 50(4):3595–3601. https://doi.org/10.1002/2013WR015116
97. Hearst MA, Dumais ST, Osuna E, Platt J, Scholkopf B (1998) Support vector machines. IEEE Intell Syst Appl 13(4):18–28. https://doi.org/10.1109/5254.708428
98. Hirschmann MM, Ghiorso MS, Davis FA, Gordon SM, Mukherjee S, Grove TL, Krawczynski M, Medard E, Till CB, Medard E, Till CB (2008) Library of Experimental Phase Relations (LEPR): a database and Web portal for experimental magmatic phase equilibria data. Geochem Geophys Geosyst 9(3), n/a-n/a. https://doi.org/10.1029/2007GC001894
99. Ho T (1998) The random subspace method for constructing decision forests. IEEE Trans Patt Anal Mach Intell 20(8):832–844. https://doi.org/10.1109/34.709601
100. Huang C, Davis L, Townshend J (2002) An assessment of support vector machines for land cover classification. Int J Remote Sens 23(4):725–749. https://doi.org/10.1080/ 01431160110040323
101. Huber PJ, Ronchetti EM (2009) Robust statistics, 2nd edn. Wiley
102. Jain A, Murty M, Flynn P (1999) Data clustering: a review. ACM Comput Surv 31(3):264–323. https://doi.org/10.1145/331499.331504
103. Jordan M, Mitchell T (2015) Machine learning: trends, perspectives, and prospects. Science 349(6245):255–260. https://doi.org/10.1126/science.aaa8415

104. Kotsiantis S (2007) Supervised machine learning: a review of classification techniques. Informatica (Ljubljana) 31(3):249–268
105. Le Maitre R (1982) Numerical petrology. Elsevier
106. Lee J, Verleysen M (2009) Quality assessment of dimensionality reduction: rank-based criteria. Neurocomputing 72(7–9):1431–1443. https://doi.org/10.1016/j.neucom.2008.12.017
107. Maronna RA, Martin RD, Yohai VJ (2006) Robust statistics: theory and methods, 2nd edn. Wiley. https://doi.org/10.1002/0470010940
108. Masotta M, Mollo S, Freda C, Gaeta M, Moore G (2013) Clinopyroxene-liquid thermometers and barometers specific to alkaline difierentiated magmas. Contribut Mineral Petrol 166(6):1545–1561. https://doi.org/10.1007/s00410-013-0927-9
109. Masotti M, Falsaperla S, Langer H, Spampinato S, Campanini R (2006) Application of support vector machine to the classification of volcanic tremor at Etna, Italy. Geophys Res Lett 33(20). https://doi.org/10.1029/2006GL027441
110. Murphy K (2012) Machine learning. MIT Press
111. Natekin A, Knoll A (2013) Gradient boosting machines, a tutorial. Front Neurorobot 7:21. https://doi.org/10.3389/fnbot.2013.00021
112. Neave DA, Bali E, Guðfinnsson GH, Halldórsson SA, Kahl M, Schmidt AS, Holtz F (2019) Clinopyroxene-Liquid Equilibria and Geothermobarometry in Natural and Experimental Tholeiites: The 2014–2015 Holuhraun Eruption. Iceland. J Petrol 60(8):1653–1680. https://doi.org/10.1093/petrology/egz042
113. Nimis P (1995) A clinopyroxene geobarometer for basaltic systems based on crystal-structure modeling. Contribut Mineral Petrol 121(2):115–125. https://doi.org/10.1007/s004100050093
114. Nimis P, Ulmer P (1998) Clinopyroxene geobarometry of magmatic rocks Part 1: an expanded structural geobarometer for anhydrous and hydrous, basic and ultrabasic systems. Contribut Mineral Petrol 133(1–2):122–135. https://doi.org/10.1007/s004100050442
115. Palettas PN (1992) Probability Plots and Modern Statistical Software. In: Page C, LePage R (eds) Computing science and statistics. Springer, New York, pp 489–491
116. Pedregosa F, Varoquaux GG, Gramfort A, Michel V, Thirion B, Grisel O, Blondel M, Prettenhofer P, Weiss R, Dubourg V, Vanderplas J, Passos A, Cournapeau D, Brucher M, Perrot M, Duchesnay É (2011) Scikit-learn: machine learning in Python. J Mach Learn Res 12(85):2825–2830
117. Petrelli M, Bizzarri R, Morgavi D, Baldanza A, Perugini D (2017) Combining machine learning techniques, microanalyses and large geochemical datasets for tephrochronological studies in complex volcanic areas: New age constraints for the Pleistocene magmatism of central Italy. Quat Geochronol 40:33–44. https://doi.org/10.1016/j.quageo.2016.12.003
118. Petrelli M, Caricchi L, Perugini D (2020) Machine learning thermo- barometry: application to clinopyroxene-bearing magmas. J Geophys Res: Solid Earth 125(9). https://doi.org/10.1029/2020JB020130
119. Petrelli M, Perugini D (2016) Solving petrological problems through machine learning: the study case of tectonic discrimination using geochemical and isotopic data. Contribut Mineral Petrol 171(10). https://doi.org/10.1007/s00410-016-1292-2
120. Petrelli M, Perugini D, Moroni B, Poli G (2003) Determination of travertine provenance from ancient buildings using self-organizing maps and fuzzy logic. Appl Artif Intell 17(8–9):885–900. https://doi.org/10.1080/713827251
121. Putirka K (2008) Introduction to minerals, incusions and volcanic processes. Rev Mineral Geochem 69:1–8. https://doi.org/10.2138/rmg.2008.69.1
122. Putirka K, Mikaelian H, Ryerson F, Shaw H (2003) New clinopyroxene-liquid thermobarometers for mafic, evolved, and volatile-bearing lava compositions, with applications to lavas from Tibet and the Snake River Plain. Idaho Am Mineral 88(10):1542–1554. https://doi.org/10.2138/am-2003-1017
123. Shai S-S, Shai B.-D (2013) Understanding machine learning: from theory to algorithms, vol. 9781107057. Cambridge University Press. https://doi.org/10.1017/CBO9781107298019
124. Shapiro SS, Wilk MB (1965) An analysis of variance test for normality (complete samples). Biometrika 52(3/4):591. https://doi.org/10.2307/2333709

125. Smola A, Schölkopf B (2004) A tutorial on support vector regression. Stat Comput 14(3):199–222. https://doi.org/10.1023/B:STCO.0000035301.49549.88
126. Stephens MA (1974) EDF statistics for goodness of fit and some comparisons. J Am Stat Assoc 69:730–737
127. Thode HC (2002) Testing for normality. CRC Press
128. Zhang Y, Haghani A (2015) A gradient boosting method to improve travel time prediction. Transport Res Part C: Emerg Technol 58:308–324. https://doi.org/10.1016/j.trc.2015.02.019
129. Zuo R, Carranza E (2011) Support vector machine: a tool for mapping mineral prospectivity. Comput Geosci 37(12):1967–1975. https://doi.org/10.1016/j.cageo.2010.09.014

9 783030 780548